CAMBRIDGE LIBRARY COLLECTION

Books of enduring scholarly value

Darwin

Two hundred years after his birth and 150 years after the publication of 'On the Origin of Species', Charles Darwin and his theories are still the focus of worldwide attention. This series offers not only works by Darwin, but also the writings of his mentors in Cambridge and elsewhere, and a survey of the impassioned scientific, philosophical and theological debates sparked by his 'dangerous idea'.

Rough Notes Taken During Some Rapid Journeys Across the Pampas and Among the Andes

Sir Francis Bond Head (1793–1875) known as 'Galloping Head', was a soldier who later served as lieutenant-governor of Upper Canada, but who was dismissed from his post when rebellion broke out there in 1837. Before this, he had tried unsuccessfully to set up a mining company in Argentina. It is from this period of his life that the characteristically entitled Rough Notes Taken During Some Rapid Journeys Across the Pampas and Among the Andes (published in 1826) were written, in a headlong and jocular style which belies the actual hardships of his journey. Part of the interest of the account today lies in the fact that Charles Darwin had read it and refers to it frequently and admiringly in his letters home as he traversed the same country six years later: 'Do you know Head's book? it gives an excellent account of the manners of this country.'

Cambridge University Press has long been a pioneer in the reissuing of out-of-print titles from its own backlist, producing digital reprints of books that are still sought after by scholars and students but could not be reprinted economically using traditional technology. The Cambridge Library Collection extends this activity to a wider range of books which are still of importance to researchers and professionals, either for the source material they contain, or as landmarks in the history of their academic discipline.

Drawing from the world-renowned collections in the Cambridge University Library, and guided by the advice of experts in each subject area, Cambridge University Press is using state-of-the-art scanning machines in its own Printing House to capture the content of each book selected for inclusion. The files are processed to give a consistently clear, crisp image, and the books finished to the high quality standard for which the Press is recognised around the world. The latest print-on-demand technology ensures that the books will remain available indefinitely, and that orders for single or multiple copies can quickly be supplied.

The Cambridge Library Collection will bring back to life books of enduring scholarly value (including out-of-copyright works originally issued by other publishers) across a wide range of disciplines in the humanities and social sciences and in science and technology.

Rough Notes Taken During Some Rapid Journeys Across the Pampas and Among the Andes

Francis Bond Head

CAMBRIDGE
UNIVERSITY PRESS

CAMBRIDGE UNIVERSITY PRESS

Cambridge, New York, Melbourne, Madrid, Cape Town, Singapore,
São Paolo, Delhi, Dubai, Tokyo

Published in the United States of America by Cambridge University Press, New York

www.cambridge.org
Information on this title: www.cambridge.org/9781108001618

© in this compilation Cambridge University Press 2009

This edition first published 1826
This digitally printed version 2009

ISBN 978-1-108-00161-8 Paperback

ROUGH NOTES

TAKEN DURING

SOME RAPID JOURNEYS

ACROSS

THE PAMPAS

AND AMONG

THE ANDES.

———

By CAPTAIN F. B. HEAD.

———

LONDON:

JOHN MURRAY, ALBEMARLE-STREET.

———

MDCCCXXVI.

INTRODUCTION.

THE sudden rise and fall, the unexpected appearance and disappearance, of so many Mining Companies, is a subject which must necessarily occupy a few lines in the future history of our country; and when the exultation of those who have gained, and the disappointment of those who have lost, are alike forgotten, the Historian who calmly relates the momentary existence of these Companies, will only inquire into the general causes of their formation, and the general causes of their failure.

That a commercial error has been committed, no one can deny; and it must also be admitted, that this error was not confined to a few individuals, or to any association of individuals, but like a contagious disease it pervaded all classes of society; and that among the lists of Shareholders in these speculations, were to be found the names of people of the first rank, character, and education in the country.

Experience has at last been purchased at a very great loss, and by it we now learn, that both the formation of these Companies, and their failure, have proceeded from one cause—our Ignorance of the country which was to be the field of the speculation. But although this must be confessed, yet let it also be remembered, that the error was accompanied by all the noble characteristics which distinguish our country.

Had we known the nature of the different countries, it would have been deemed imprudent to have forwarded to them expensive machinery, to have given liberal salaries to every individual connected with the speculation, to have invited the Natives to share the profits, to have intrusted the Capital to solitary individuals, &c. Still had the Foundation been good, the Building was nobly planned, and it was undeniably the act and the invention of a country teeming with energy, enterprise, liberality, unsuspecting confidence, and capital.

Without lamenting over losses which are now irrecoverable, it is only necessary to keep in mind that the *Cause* which produced them still exists, and that we are still in ignorance of the countries in which our money lies buried. Many of the individuals who had charge of the different Companies, had undoubtedly opportunities of making important observations, and from them valuable data will probably be obtained.

I myself had the sole management of one of these Companies; but, from particular circumstances, it will be proper to show that, excepting for my Reports, I had little time or opportunity to make any memoranda beyond those of the most trifling description of personal narrative.

I was on duty at Edinburgh, in the corps of Engineers, when it was proposed to me to take charge of an Association, the object of which was to work the Gold and Silver Mines of the Provinces of Rio de la Plata; and, accordingly, at a very few days' notice, I sailed from Falmouth, and landed at Buenos Aires about a week after the Cornish Miners had arrived there.

Accompanied by two highly respectable Captains of the Cornish Mines, a French Assayer, who had been brought up by the celebrated Vauquelin, a Surveyor, and three miners, I proceeded across the great plains of the Pampas to the Gold Mines of San Luis, and from thence to the Silver Mines of Uspallata which are beyond Mendoza, about a thousand miles from Buenos Aires.

I then left my party at Mendoza, and from the Mines I rode back again to Buenos Aires by myself, performing the distance in eight days. I there unexpectedly received letters which made it necessary for me to go immediately to Chili, and I accordingly again crossed the Pampas, and, joining my party at Mendoza, we went over the Andes to Santiago, and from thence, without any de-

lay, we went together in different directions
about twelve hundred miles, to inspect gold
and silver mines; and on the night that I
concluded my report on the last mine, we
again set off to recross the Cordillera, and
leaving my party in the plains, I rode across
the Pampas to Buenos Aires, and as soon
as I arrived there I dismissed a proportion
of the miners, and returned with the rest to
England.

The sole object of my journeys was to
inspect certain mines.   We went to the bottom of them all, and, assisted by the individuals who accompanied me, I made, to
the best of my ability, a circumstantial report on each.   As the miners were remaining idle and without employment at Buenos
Aires, it was highly desirable that I should
go from place to place as rapidly as possible,

and for upwards of six thousand miles I can truly declare that I was riding against Time.

The fatigue of such long journeys, exposed to the burning sun of summer, was very great, and particularly in Chili, because, in visiting mines in the Andes, we were subjected to such sudden changes of climate, that we were occasionally overpowered by the sun in the morning, while at night we had to sleep upon one hundred and twenty feet of snow; for almost the whole time we slept out on the ground, chiefly subsisting upon beef and water.

The reports which I collected, and the result of the communications which I officially had with the Ministers, Governors, and other individuals concerning the mines, I do not feel inclined to publish; because as the mines which I visited almost all belong to private individuals, and are now for sale, it might be considered a violation of the attentions which I often received, to state unnecessarily the dimensions, contents, or the assay of their lodes, although the climate and the general features of the country are, of course, public property.

During my journeys I kept no regular journal, for the country I visited was either a boundless plain, or desert mountains; but I occasionally made a few rough notes, describing anything which interested or amused me.

These notes were written under great variety of circumstances, sometimes when I was tired, sometimes when I was refreshed, sometimes with a bottle of wine before me, and sometimes with a cow's-horn filled with dirty brackish water, and a few were written on board the packet.

They were only made to amuse my mind
under a weight of responsibility to which
it had never been accustomed, and therefore they are necessarily in that incoherent,
unconnected state which makes them, I am
fully aware, but little suited to meet the
critical eye of the public; still as it has been
my misfortune to see the failure of an English Association—to witness the loss it has
sustained—and for a few moments at Buenos
Aires and Monte Video to stand upon spots
where we have lost what no money can re-
pay us ; as I feel persuaded that these
failures have proceeded from our ignorance
of the country, I have resolved upon throwing before the public the few memoranda I
possess, and although I am conscious that
they are of too trifling a nature to throw
much light upon the subject, yet they may,
perhaps, assist in making the " darkness

visible," and I trust that the rough, un-
polished state in which they appear will
at least be a proof that I have no other
object.

Lower Grosvenor-Street,
September 1, 1826.

DESCRIPTIVE OUTLINE

OF THE

PAMPAS,

&c. &c.

THE mountains of the Andes run about North and South through the whole of South America, and they are consequently nearly parallel to the two shores of the Pacific and the Atlantic Oceans, dividing the country between them into two unequal parts, each bounded by an Ocean and by the Cordillera.

It would at first be expected that these twin countries, separated only by a range of mountains, should have a great resemblance to each other; but variety is the attribute of Omnipotence, and nature has granted to these two countries a difference of climate and geological construction which is very remarkable.

From the tops of the Andes she supplies both

of them with water; by the gradual melting of the snow they are both irrigated exactly in proportion to their wants; and vegetation, instead of being exhausted by the burning sun of summer, is thus nourished and supported by the very heat which threatened to destroy it.

The water, however, which flows through Chili towards the Pacific, is confined in its whole course, and forces its way through a country as mountainous as the highlands of Scotland or Switzerland. The water which descends from the east side of the Cordillera meanders through a vast plain nine hundred miles in breadth; and at the top of the Andes, it is singular to observe on the right and left the snow of one storm, part of which is decreed to rush into the Pacific, while the other is to add to the distant waves of the Atlantic.

The great plain, or Pampas, on the east of the Cordillera, is about nine hundred miles in breadth, and the part which I have visited, though under the same latitude, is divided into regions of different climate and produce. On leaving Buenos Aires, the first of these regions is covered for one hundred and eighty miles with clover and thistles;

the second region, which extends for four hundred and fifty miles, produces long grass; and the third region, which reaches the base of the Cordillera, is a grove of low trees and shrubs. The second and third of these regions have nearly the same appearance throughout the year, for the trees and shrubs are evergreens, and the immense plain of grass only changes its colour from green to brown; but the first region varies with the four seasons of the year in a most extraordinary manner. In winter, the leaves of the thistles are large and luxuriant, and the whole surface of the country has the rough appearance of a turnip-field. The clover in this season is extremely rich and strong; and the sight of the wild cattle grazing in full liberty on such pasture is very beautiful. In spring, the clover has vanished, the leaves of the thistles have extended along the ground, and the country still looks like a rough crop of turnips. In less than a month the change is most extraordinary; the whole region becomes a luxuriant wood of enormous thistles, which have suddenly shot up to a height of ten or eleven feet, and are all in full bloom. The road or path is hemmed in on both

sides; the view is completely obstructed; not an
animal is to be seen; and the stems of the thistles
are so close to each other, and so strong, that,
independent of the prickles with which they are
armed, they form an impenetrable barrier.   The
sudden growth of these plants is quite astonishing; and though it would be an unusual misfortune in military history, yet it is really possible,
that an invading army, unacquainted with this
country, might be imprisoned by these thistles before they had time to escape from them.   The summer is not over before the scene undergoes another
rapid change: the thistles suddenly lose their sap
and verdure, their heads droop, the leaves shrink
and fade, the stems become black and dead, and
they remain rattling with the breeze one against
another, until the violence of the pampero or hurricane levels them with the ground, where they
rapidly decompose and disappear—the clover rushes
up, and the scene is again verdant.

Although a few individuals are either scattered
along the path, which traverses these vast plains,
or are living together in small groups, yet the
general state of the country is the same as it has

been since the first year of its creation. The whole country bears the noble stamp of an Omnipotent Creator, and it is impossible for any one to ride through it, without feelings which it is very pleasing to entertain; for although in all countries " the heavens declare the glory of God, and the firmament sheweth his handy work," yet the surface of populous countries affords generally the insipid produce of man's labour; it is an easy error to consider that he who has tilled the ground, and has sown the seed, is the author of his crop, and, therefore, those who are accustomed to see the confused produce, which in populous and cultivated countries is the effect of leaving ground to itself, are at first surprised in the Pampas, to observe the regularity and beauty of the vegetable world when left to the wise arrangements of Nature.

The vast region of grass in the Pampas for four hundred and fifty miles is without a weed, and the region of wood is equally extraordinary. The trees are not crowded, but in their growth such beautiful order is observed, that one may gallop between them in every direction. The young trees are rising up, others are flourishing in full vigour, and it is for some time that one looks in vain for those which in the great system of succession must necessarily somewhere or other be sinking towards decay. They are at last discovered, but their fate is not allowed to disfigure the general cheerfulness of the scene, and they are seen enjoying what may literally be termed a green old age. The extremities of their branches break off as they die, and when nothing is left but the hollow trunk, it is still covered with twigs and leaves, and at last is gradually concealed from view by the young shoot, which, born under the shelter of its branches, now rises rapidly above it, and conceals its decay. A few places are met with which have been burnt by accident, and the black desolate spot, covered with the charred trunks of trees, resembles a scene in the human world of pestilence or war. But the fire is scarcely extinct, when the surrounding trees all seem to spread their branches towards each other, and young shrubs are seen rising out of the ground, while the sapless trunks are evidently mouldering into dust.

The rivers all preserve their course, and the whole country is in such beautiful order, that if cities and millions of inhabitants could suddenly be planted at proper intervals and situations, the people would have nothing to do but to drive out their cattle to graze, and, without any previous preparation, to plough whatever quantity of ground their wants might require.

The climate of the Pampas is subject to a great difference of temperature in winter and summer, though the gradual changes are very regular. The winter is about as cold as our month of November, and the ground at sunrise is always covered with white frost, but the ice is seldom more than one-tenth of an inch thick. In summer the sun is very oppressively hot*, and its force is acknowledged by every living animal. The wild horses and cattle are evidently exhausted by it, and the *siesta* seems to be a repose which is natural and necessary to all. The middle of the day is not a moment for work, and as the mornings are cool, the latter are evidently best adapted for labour, and the former for repose.

* I have twice ridden across the Morea, which lies nearly in the same latitude (north) as the path across the Pampas, and I think the climate of the latter is hotter than the Morea, Sicily, Malta, or Gibraltar, in summer, and colder in winter.

The difference between the atmosphere of Mendoza, St. Lewis, and Buenos Aires, which are all nearly under the same latitude, is very extraordinary: in the two former, or in the regions of wood and grass, the air is extremely dry; there is no dew at night; in the hottest weather there is apparently very little perspiration, and the dead animals lie on the plain dried up in their skins, so that occasionally I have at first scarcely been able to determine whether they were alive or dead.   But in the province of Buenos Aires, or in the region of thistles and clover, vegetation clearly announces the humidity of the climate.   In sleeping out at night I have found my poncho (or rug) nearly wet through with the dew, and my boots so damp that I could scarcely draw them on.   The dead animals on the plain are in a rapid state of putrefaction. On arriving at Buenos Aires, the walls of the houses are so damp that it is cheerless to enter them; and sugar, as also all deliquescent salts, are there found nearly dissolved. This dampness, however, does not appear to be unhealthy.   The Gauchos and even travellers sleep on the ground, and the inhabitants of Buenos Aires live in their damp houses without complaining of rheumatism, or being at all subject to cold ; and they certainly have the appearance of being rather more robust and healthy than those who live in the drier regions. However, the whole of the Pampas may be said to enjoy as beautiful and as salubrious an atmosphere as the most healthy parts of Greece and Italy, and without being subject to malaria.

The only irregularity in the climate is the pampero or south-west wind, which, generated by the cold air of the Andes, rushes over these vast plains with a velocity and a violence which it is almost impossible to withstand. But this rapid circulation of the atmosphere has very beneficial effects, and the weather, after one of these tempests, is always particularly healthy and agreeable.

The south part of the Pampas is inhabited by the Pampas Indians, who have no fixed abode, but wander from place to place as the herbage around them becomes consumed by their cattle. The north part of the Pampas, and the rest of the Provinces of the Rio de la Plata, are inhabited by a few straggling individuals, and a few small groups of people, who live together only because they were

born together. Their history is really very
curious.

As soon as by the fall of the Spaniards their independence was established, and they became free,
the attention of many individuals of the Provinces
of La Plata was directed towards the due constitution of governments which might maintain the freedom that was gained, encourage population, and
gradually embellish the surface of a most interesting
and beautiful country with the arts, manufactures,
and sciences, which had hitherto been denied it;
but the singular situation of the country presented
very serious difficulties.

Although immense regions of rich land lay uncultivated and unowned, yet something had been
done. Small towns and establishments (originally
chosen for mining purposes,) five hundred and
seven hundred miles distant from one another, were
thinly scattered over this vast extent of country;
and thus a skeleton map of civilization had been
traced, which the narrow interests of every individual naturally supported.

But although a foundation was thus laid, the
building plan of the Spaniards was missing. It

had been destroyed in the war, and all that was known of it was, that it was for purposes which were not applicable to the great political system which should now be adopted.

It was soon perceived that the Provinces of the Rio de la Plata were without a harbour; that the town of Buenos Aires was badly situated; and as the narrow policy of Spain had forbidden the planting of the olive and the grape, the spots which were best adapted to the natural produce of the country had been neglected : while for mining, and other purposes connected with the Spanish system, towns had been built in the most remote and impracticable situations; and men found themselves living together in groups they knew not why, under circumstances which threw a damp over exertion, and under difficulties which it appeared hopeless to encounter.

Their situation was, and still is, very lamentable. The climate easily affords them the few necessaries of life. Away from all practicable communication with the civilized world, they are unable to partake of the improvements of the age, or to shake off the errors and the disadvantages of a bad political education. They have not the moral means of improving their country, or of being improved by it; and oppressed by these and other disadvantages, they naturally yield to habits of indolence and inactivity. The Town, or rather the secluded Village, in which they live, is generally the seat of government of the province, and but too often affords a sad political picture.

People who, although they are now free, were brought up under the dark tyranny of the Spanish government, with the narrow prejudices which even in populous countries exist among the inhabitants of small communities, and with little or no education, are called upon to elect a governor, and to establish a junta, to regulate the affairs of their own province, and to send a deputy to a distant national assembly at Buenos Aires. The consequence (as I have witnessed) is what might naturally be expected. The election of the governor is seldom unanimous, and he is scarcely seated before he is overturned, in a manner which, to one accustomed to governments on a larger scale, appears childish and ridiculous.

In more than one province the governor is ex-

ceedingly tyrannical: in the others, the governor
and the junta appear to act for the interests of
their own province; but their funds are so small,
and the internal jealousies they have to encounter
so great, that they meet with continual difficulties;
and with respect to acting for the national interest,
the thing is impossible. How can it be expected
that people of very slender incomes, and in very
small insulated societies, will forget their own narrow interests for the general welfare of their country? It is really against Nature, for what is po-
litically termed their country, is such an immense
space, that it must necessarily become the future
seat of many different communities of men; and if
these communities, however enlightened they may
become, will never be able to conquer that feeling
which endears them to their homes, or the centri-
fugal prejudice with which they view their neigh-
bours, how can it be expected that a feeble government and a few inhabitants can do what civilization
has not yet been able to perform; or that the
political infant will not betray those frailties which
his manhood will be incapable of overcoming. And
the fact is, that each Province does view its neigh-

bouring one with jealousy, and as I have travelled through the country, 1 have invariably found that *mala gente* is the general appellation which the people give to those of the adjoining province, and that they, as well as the inhabitants of the towns, are all jealous of the power and influence of the town of Buenos Aires; and when it is explained, that the policy of Buenos Aires is to break the power of the monks and priests, and that these people have still very great influence in most of the distant provinces, and that the maritime interest of Buenos Aires is necessarily often at variance with that of the inland provinces, it will be perceived how forcibly this jealousy is likely to act.

The situation of the Gaucho is naturally independent of the political troubles which engross the attention of the inhabitants of the towns. The population or number of these Gauchos is very small, and at great distances from each other: they are scattered here and there over the face of the country. Many of these people are descended from the best families in Spain; they possess good-manners, and often very noble sentiments: the life they lead is very interesting—they generally inhabit the hut in which they were born, and in which their fathers and grandfathers lived before them, although it appears to a stranger to possess few of the allurements of *dulce domum*. The huts are built in the same simple form; for although luxury has ten thousand plans and elevations for the frail abode of its more frail tenant, yet the hut in all countries is the same, and therefore there is no difference between that of the South American Gaucho, and the Highlander of Scotland, excepting that the former is built of mud, and covered with long yellow grass, while the other is formed of stones, and thatched with heather. The materials of both are the immediate produce of the soil, and both are so blended in colour with the face of the country, that it is often difficult to distinguish them; and as the pace at which one gallops in South America is rapid, and the country flat, one scarcely discovers the dwelling before one is at the door. The corral is about fifty or one hundred yards from the hut, and is a circle of about thirty yards in diameter, enclosed by a number of strong rough posts, the ends of which are struck into the ground. Upon these posts are generally a number of idle-looking vultures or hawks *, and the ground around the hut and corral is covered with bones and carcasses of horses, bullocks' horns, wool, &c., which give it the smell and appearance of an ill-kept dog-kennel in England.

The hut consists generally of one room, in which all the family live, boys, girls, men, women, and children, all huddled together. The kitchen is a detached shed a few yards off: there are always holes, both in the walls and in the roof of the hut, which one at first considers as singular marks of the indolence of the people. In the summer this

---

* The hawks are very tame, and they are seldom to be seen except at the huts; but occasionally they have followed me for many leagues, keeping just before me, and with their round black eyes gazing intently on my face, which I fancied attracted their notice from being burnt by the sun, and I literally often thought they were a little inclined to taste it. They are constantly in the habit of attacking the horses and mules who have sore backs; and I have often observed these birds hovering about six inches above them. It is curious to compare the countenance of the two animals. The hawk, with his head bent downwards, and his eye earnestly fixed upon the wound: the mule with his back crouched down, his ears lying back, whisking his tail, afraid to eat, and apparently not knowing whether to rear or kick.

abode is so filled with fleas and binchucas, (which are bugs as large as black beetles,) that the whole family sleep on the ground in front of their dwelling; and when the traveller arrives at night, and after unsaddling his horse walks among this sleeping community, he may place the saddle or recado on which he is to sleep close to the companion most suited to his fancy:—an admirer of innocence may lie down by the side of a sleeping infant; a melancholy man may slumber near an old black woman; and one who admires the fairer beauties of creation, may very demurely lay his head on his saddle, within a few inches of the idol he adores. However, there is nothing to assist the judgment but the bare feet and ancles of all the slumbering group, for their heads and bodies are covered and disguised by the skin and poncho which cover them.

In winter the people sleep in the hut, and the scene is a very singular one. As soon as the traveller's supper is ready, the great iron spit on which the beef has been roasted is brought into the hut, and the point is struck into the ground: the Gaucho then offers his guest the skeleton of a horse's head, and he and several of the family, on similar seats, sit round the spit, from which with their long knives they cut very large mouthfuls*. The hut is lighted by a feeble lamp, made of bullock's tallow; and it is warmed by a fire of charcoal: on the walls of the hut are hung, upon bones, two or three bridles and spurs, and several lassos and balls: on the ground are several dark-looking heaps, which one can never clearly distinguish; on sitting down upon these when tired, I have often heard a child scream underneath me, and have occasionally been mildly asked by a young woman, what I wanted?—at other times up has jumped an immense dog! While I was once warming my hands at the fire of charcoal, seated on a horse's head, looking at the black roof in a reverie, and fancying I was quite by myself, I felt something touch me, and saw two naked black children leaning over the charcoal in the attitude of two toads: they had crept out from under some of

* When first I lived with the Gauchos, I could not conceive how they possibly managed to eat so quickly meat which I found so unusually tough, but an old Gaucho told me it was because I did not know what parts to select, and he immediately cut me a large piece which was quite tender. I always afterwards begged the Gauchos to help me, and they generally smiled at my having discovered the secret.

the ponchos, and I afterwards found that many other persons, as well some as hens sitting upon eggs, were also in the hut. In sleeping in these huts, the cock has often hopped upon my back to crow in the morning; however, as soon as it is daylight, everybody gets up.

The life of the Gaucho is very interesting, and resembles that beautiful description which Horace gives of the progress of a young eagle:—

> Olim juventas et patrius vigor
> Nido laborum propulit inscium,
> Vernique jam nimbis remotis
> Insolitos docuêre nisus
> Venti paventem; mox in ovilia
> Demisit hostem vividus impetus,
> Nunc in reluctantes dracones
> Egit amor dapis, atque pugnæ.

Born in the rude hut, the infant Gaucho receives little attention, but is left to swing from the roof in a bullock's hide, the corners of which are drawn towards each other by four strips of hide. In the first year of his life he crawls about without clothes, and I have more than once seen a mother give a child of this age a sharp knife, a foot long, to play with. As soon as he walks, his infantine amuse-

ments are those which prepare him for the occupations of his future life: with a lasso made of twine
he tries to catch little birds, or the dogs, as they
walk in and out of the hut. By the time he is four
years old he is on horseback, and immediately
becomes useful by assisting to drive the cattle into
the corral. The manner in which these children
ride is quite extraordinary: if a horse tries to
escape from the flock which are driven towards the
corral, I have frequently seen a child pursue him,
overtake him, and then bring him back, flogging
him the whole way; in vain the creature tries to
dodge and escape from him, for the child turns
with him, and always keeps close to him ; and it is
a curious fact, which I have often observed, that a
mounted horse is always able to overtake a loose
one.

His amusements and his occupations soon become
more manly—careless of the biscacheros (the holes
of an animal called the biscacho) which undermine
the plains, and which are very dangerous, he gallops after the ostrich, the gama, the lion, and the
tiger; he catches them with his balls : and with his
lasso he daily assists in catching the wild cattle, and

in dragging them to the hut either for slaughter, or to be marked. He breaks in the young horses in the manner which I have described, and in these occupations is often away from his hut many days, changing his horse as soon as the animal is tired, and sleeping on the ground. As his constant food is beef and water, his constitution is so strong that he is able to endure great fatigue; and the distances he will ride, and the number of hours that he will remain on horseback, would hardly be credited. The unrestrained freedom of such a life he fully appreciates; and, unacquainted with subjection of any sort, his mind is often filled with sentiments of liberty which are as noble as they are harmless, although they of course partake of the wild habits of his life. Vain is the endeavour to explain to him the luxuries and blessings of a more civilized life; his ideas are, that the noblest effort of man is to raise himself off the ground and ride instead of walk—that no rich garments or variety of food can atone for the want of a horse—and that the print of the human foot on the ground is in his mind the symbol of uncivilization.

The Gaucho has by many people been accused of

indolence; those who visit his hut find him at the door with his arms folded, and his poncho thrown over his left shoulder like a Spanish cloak; his hut is in holes, and would evidently be made more comfortable by a few hours' labour: in a beautiful climate, he is without fruit or vegetables; surrounded by cattle, he is often without milk; he lives without bread, and he has no food but beef and water, and therefore those who contrast his life with that of the English peasant accuse him of indolence; but the comparison is inapplicable, and the accusation unjust; and any one who will live with the Gaucho, and will follow him through his exertions, will find that he is any thing but indolent, and his surprise will be that he is able to continue a life of so much fatigue. It is true that the Gaucho has no luxuries, but the great feature of his character is, that he is a person without wants: accustomed constantly to live in the open air, and to sleep on the ground, he does not consider that a few holes in his hut deprive it of its comfort. It is not that he does not like the taste of milk, but he prefers being without it to the every-day occupation of going in search of it. He might, it is

trüe, make cheese, and sell it for money, but if he
has got a good saddle and good spurs, he does not
consider that money has much value: in fact, he is
contented with his lot; and when one reflects that,
in the increasing series of human luxuries, there is
no point that produces contentment, one cannot but
feel that there is perhaps as much philosophy as
folly in the Gaucho's determination to exist without
wants; and the life he leads is certainly more noble
than if he was slaving from morning till night to
get other food for his body or other garments to
cover it. It is true he is of little service to the
great cause of civilization, which it is the duty of
every rational being to promote; but an humble individual, living by himself in a boundless plain,
cannot introduce into the vast uninhabited regions
which surround him either arts or sciences: he may,
therefore, without blame be permitted to leave them
as he found them, and as they must remain, until
population, which will create wants, devises the
means of supplying them.

The character of the Gaucho is often very
estimable; he is always hospitable—at his hut
the traveller will always find a friendly welcome,

and he will often be received with a natural dignity
of manner which is very remarkable, and which he
scarcely expects to meet with in such a miserable-
looking hovel.   On entering the hut, the Gaucho
has constantly risen to offer me his seat, which I
have declined, and many compliments and bows
have passed, until I have accepted his offer, which
is the skeleton of a horse's head.   It is curious to
see them invariably take off their hats to each other
as they enter into a room which has no window, a
bullock's hide for a door, and but little roof.

The habits of the women are very curious: they
have literally nothing to do; the great plains
which surround them offer them no motive to walk,
they seldom ride, and *their* lives certainly are very
indolent and inactive.   They have all, however,
families, whether married or not; and once when I
inquired of a young woman employed in nursing
a very pretty child, who was the father of the
" creatura," she replied, Quien sabe?

The religion which is professed throughout the
provinces of the Rio de la Plata is the Roman
Catholic, but it is very different in different places.
During the reign of the Spaniards, the monks and

priests had everywhere very great influence; and
the dimensions of the churches at Buenos Aires,
Lucan, Mendoza, &c., show the power and riches
they possessed, and the greedy ambition which
governed them. It is a sad picture to see a number of small, wretched-looking huts surrounding a
church whose haughty elevation is altogether inapplicable to the humility of the Christian religion;
and one cannot help comparing it with the quiet village church of England, whose exterior and interior tends rather to humble the feelings of the arro-
gant and proud, while to the peasant it has the
cheerful appearance of his own home. When it
is considered that the churches in South America
were principally built for the conversion of the
Indians to the Christian faith, it is melancholy to
think that the priests should have attempted, by the
pomp of their temples, and by the mummery of
candles, and pictures, and images, to have done
what by reason, and kindness, and humility, would
surely have been better performed. But their secret
object was to extort money; and as it is always
easier to attract a crowd of people by bad passions
than by good, they therefore made their temples as

attractive as possible, and men were called to see and to admire, instead of to listen and to reflect.

The power of the priests and monks has changed very much since the revolution. At Buenos Aires most of the convents have been suppressed, and the general wish of almost all parties is to suppress the remainder. Occasionally, an old mendicant friar is seen, dressed in grey sackcloth, and covered with dirt; but as he walks through the street, looking on the ground, his emaciated cheek and sunken eye show that his power is crushed, and his influence gone. The churches have lost their plate, the candles are yellow, the pictures are bad, and the images are dressed in coarse English cotton. On great days, the ladies of Buenos Aires, dressed in their best clothes, are seen going to church, followed by a black child, in yellow or green livery, who carries in his arms an English hearth-rug, always of the most brilliant colours, on which the lady kneels, with the black child behind her; but generally the churches are deserted, and nobody is to be seen in them but a decrepid old woman or two, whispering into the chinks of the confessional box. The sad consequence of all this is, that at Buenos

Aires there is very little religion at all. At Mendoza there are several people who wish to put down the priests; still, however, they have evidently considerable power.

Once a year the men and women are called upon to live for nine days in a sort of barrack, which, as a great favour, I was allowed to visit. It is filled with little cells, and the men and women, at different times, are literally shut up in these holes, to fast and whip themselves. I asked several people seriously whether this punishment was *bonâ fide* performed, and they assured me that most of them whipped themselves till they brought blood. One day, I was talking very earnestly to a person at Mendoza, at the hotel, when a poor-looking monk arrived with a little image surrounded with flowers: this image my friend was obliged to kiss, and the monk then took it to every individual in the hotel— to the landlord, his servants, and even to the black cook, who all kissed it, and then of course paid for the honour. The cook gave the monk two eggs.

The priests at Mendoza lead a dissolute life; most of them have families, and several live openly with their children. Their principal amusement, however, odd as it may sound, is cockfighting every Thursday and Sunday. I was riding one Sunday when I first discovered their arena, and got off my horse to look at it. It was crowded with priests, who had each a fighting-cock under his arm; and it was surprising to see how earnest and yet how long they were in making their bets. I stayed there more than an hour, during which time the cocks were often upon the point of fighting, but the bet was not settled. Besides the priests, there were a number of little dirty boys, and one pretty-looking girl present. While they were arranging their bets, the boys began to play, so the judge instantly ordered all those who had no cocks to go out of the arena; upon which the poor girl and all the little boys were immediately turned out.

I soon got tired of the scene; but before I left them, I could not help thinking what an odd sight it was, and how justly shocked people in England would be to see a large body of clergymen fighting cocks upon a Sunday.

At St. Juan the priests have rather more power than at Mendoza, and this they shewed the other day, by taking the governor prisoner while in bed, and by burning, by the hands of the jailer, on the Plaza, the Carta de Mayo, which, to encourage the settlement of the English in this province, had lately granted to strangers religious toleration. In the other provinces the priests have more or less power, according to their abilities, and generally according to their greater or less communication with Buenos Aires.

The religion of the Gaucho is necessarily more simple than in the town, as his situation places him out of the reach of the priests. In almost all the huts there is a small image or picture, and the Gauchos have sometimes a small cross round their necks. In order that their children should be baptized, they carry them on horseback to the nearest church, and I believe the dead are generally thrown across a horse and buried in consecrated ground: though the courier and postilion who were murdered, and whose funeral service I attended, were buried in the ruins of an old hut in the middle of the Plain of Sta. Fé. When a marriage is contracted, the young Gaucho takes his bride behind him on his horse, and in the course of a few days they can generally get to a church.

THE TOWN OF BUENOS AIRES

is far from being an agreeable residence for those who are accustomed to English comforts. The water is extremely impure, scarce, and consequently expensive. The town is badly paved and dirty, and the houses are the most comfortless abodes I ever entered. The walls, from the climate, are damp, mouldy, and discoloured. The floors are badly paved with bricks, which are generally cracked, and often in holes. The roofs have no ceiling, and the families have no idea of warming themselves except by huddling round a fire of charcoal, which is put outside the door until the carbonic acid gas has rolled away.

Some of the principal families at Buenos Aires furnish their rooms in a very expensive, but comfortless manner: they put down upon the brick floor a brilliant Brussels carpet, hang a lustre from the rafters, and place against the damp wall, which they whitewash, a number of tawdry North American chairs. They get an English piano-forte, and some marble vases, but they have no idea of grouping their furniture into a comfortable form: the ladies sit with their backs against the walls without any apparent means of employing themselves; and when a stranger calls upon them, he is much surprised to find that they have the uncourteous custom of never rising from their chairs. I had no time to enter into any society at Buenos Aires, and the rooms looked so comfortless, that, to tell the truth, I had little inclination. The society of Buenos Aires is composed of English and French merchants, with a German or two. The foreign merchants are generally the agents of European houses; and as the customs of the Spanish South Americans, their food, and the hours at which they eat it, are different from those of the English and French, there does not appear to be much communication between them.

At Buenos Aires the men and women are rarely seen walking together; at the theatre they are completely separated; and it is cheerless to see all the ladies sitting together in the boxes, while the men are in the pit,—slaves, common sailors, soldiers, and merchants, all members of the same republic.

The town is furnished with provisions by the
Gauchos in a manner that shews a great want of
attention to those arrangements which are generally
met with in civilized communities. Milk, eggs,
fruit, vegetables, and beef are brought into the town
by individuals at a gallop*, and they are only to be
had when they choose to bring them. The articles
of life are brought together without due arrangement, and the consequence is, that (excepting beef)
they are dearer than in London, and sometimes
are not to be had at all. I happened to leave
Buenos Aires just as the fig-season was over, and
though it was the middle of summer, no fruit was
to be had: the towns-people seemed to be quite satis-

---

* One of the most striking pictures in and near Buenos Aires
is the young Gaucho who brings milk. The milk is carried in
six or eight large earthen bottles, which hang on each side of
the saddle. There is seldom room for the boy's legs, and he
therefore generally turns his feet up behind him on the saddle,
and sits like a frog. One meets these boys in squads of four or
five, and the manner in which they gallop in their red cloth
caps, with their scarlet ponchos flying behind them, has a singular appearance. The butchers' shops are covered carts, which
are not very agreeable objects. The beef, mangled in a most
shocking manner, is swinging about; and I have constantly seen
a large piece tied by a strip of hide to the tail of the cart, and
dragged along the ground, with a dog trying to tear it.

fied with this reason, and I could not persuade them that some one should arrange a constant supply and succession of fruits, and not leave it entirely to the Gaucho. But the same want of arrangement exists in all instances. If one has been taken out to dinner in a carriage, and in the evening ventures to inquire why it has not arrived, the answer is that it is raining, and that those who let carriages will not allow them to go out if it rains.

During the short time I was at Buenos Aires, I lived in a house out of the town, which was opposite the English burying-ground, and very near the place where the cattle were killed. This latter spot was about four or five acres, and was altogether devoid of pasture; at one end of it there was a large corral enclosed by rough stakes, and divided into a number of pens, each of which had a separate gate. These *cells* were always full of cattle doomed for slaughter. I several times had occasion to ride over this field, and it was curious to see its different appearances. In passing it in the day or evening, no human being was to be seen : the cattle up to their knees in mud, and with nothing to eat, were standing in the sun, occasionally low-

D

ing, or rather roaring to each other. The ground
in every direction was covered with groups of large
white gulls, some of which were earnestly pecking
at the slops of blood which they had surrounded,
while others were standing upon their tip-toes, and
flapping their wings as if to recover their appetite.
Each slop of blood was the spot where a bullock
had died; it was all that was left of his history,
and pigs and gulls were rapidly consuming it.
Early in the morning no blood was to be seen; a
number of horses, with the lassos hanging to their
saddles, were standing in groups apparently asleep:
the mataderos were either sitting or lying on the
ground close to the stakes of the corral, and
smoking segars; while the cattle, without metaphor,
were waiting until the last hour of their existence
should strike; for as soon as the clock of the Recolata struck, the men all vaulted on their horses,
the gates of all the cells were opened, and in a
very few seconds, there was a scene of apparent
confusion which it is quite impossible to describe.
Every man had a wild bullock at the end of his
lasso; some of these animals were running away
from the horses, and some were running at them;

many were roaring, some were hamstrung, and running about on their stumps; some were killed and skinned, while occasionally one would break the lasso. The horse would often fall upon his rider, and the bullock endeavour to regain his liberty, until the horsemen at full speed caught him with the lasso, tripping him off the ground in a manner that might apparently break every bone in his body. I was more than once in the middle of this odd scene, and was really sometimes obliged to gallop for my life, without exactly knowing where to go, for it was often Scylla and Charybdis.

I was one day going home from this scene when I saw a man on foot select a very large pig from a herd, and throw a lasso over his neck; he pulled it with all his strength, but the pig had no idea of obeying the summons: in an instant a little child rode up, and very quietly taking the end of the lasso from the man, he lifted up the sheep-skin which covered the saddle, fixed the lasso to the ring which is there made for it, and then instantly set off at a gallop: never did any one see an obstinate animal so completely conquered! With his tail pointing to the ground, hanging back, and with his four feet all scratching along the ground like the teeth of a harrow, he followed the boy evidently altogether against his will; and the sight was so strange, that I instantly galloped after the pig, to watch his countenance. He was as obstinate as ever until the lasso choked him, and he then fainted, and fell on his side. The boy dragged him in this state, at a gallop, more than three-quarters of a mile over hard rough ground, and at last suddenly stopped, and jumping off his horse, began to unloose the lasso:—" Sta muerto !" (he is dead,) said I to the boy, really sorry for the pig's fate. " Sta vivo !" exclaimed the child, as he vaulted on his horse, and galloped away. I watched the pig for some time, and was observing the blood on his nose, when, to my great surprise, he began to kick his hind leg: he then opened his mouth, and at last his eyes; and after he had looked about him, a little like Clarence after his dream, he got up, and very leisurely walked to a herd of ten or twelve pigs of about the same size as himself, who were about twenty yards off. I slowly followed him, and when I came to the herd, I saw they had every one of them bloody noses.

The house which I had near Buenos Aires was not only opposite the English burying-ground, but on the road to the Recolata, which was the great burying-place for the town; about half-a-dozen funerals passed my window every day, and during the few days I was at Buenos Aires, I scarcely ever rode into the town without meeting one.

Although the manners, customs, amusements, and fashions of different nations are constantly changing, and are generally different in different climates, yet one would at first expect that so simple an act as that of consigning to its narrow bed the body of a dead man would, in all countries and in all places, be the same,—but though death is the same, funerals are very different. In the old world, how often does the folly and vanity and vexation of spirit in which a man has lived accompany him to the tomb; and how often are the good feelings of the living overpowered by the vain pomp and ostentation which mocks the burial of the dead. In South America, the picture is a very different one, and certainly the way in which the people were buried at Buenos Aires appeared more strange to my eyes than any of the customs of the place. Of late years, a few of the principal people have been buried in coffins, but generally the dead are called for by a hack hearse, in which there is a fixed coffin, into which they are put, when away the man gallops with the corpse, and leaves it in the vestibule of the Recolata. There is a small vehicle for children, which I really thought was a mountebank's cart; it was a light open tray, on wheels painted white, with light blue silk curtains, and driven at a gallop by a lad dressed in scarlet, with an enormous plume of white feathers in his hat. As I was riding home one day, I was over-taken by this cart, (without its curtains, &c.) in which there was the corpse of a black boy nearly naked. I galloped along with it for some dis-tance; the boy, from the rapid motion of the carriage, was dancing sometimes on his back and sometimes on his face; occasionally his arm or leg would get through the bar of the tray, and two or three times I really thought the child would have been out of the tray altogether. The bodies of the rich were generally attended by their friends; but the carriages with four people in each were seldom able to go as fast as the hearse.

I went one day to the Recolata, and just as I got there, the little hearse drove up to the gate. The man who had charge of the burial-place received from the driver a ticket, which he read, and put into his pocket; the driver then got into the tray, and taking out a dead infant of about eight months old, he gave it to the man, who carried it swinging by one of its arms into the square-walled burial-ground, and I followed him. He went to a spot about ten yards from the corner, and then, without putting his foot upon the spade, or at all lifting up the ground, he scratched a place not so deep as the furrow of a plough. While he was doing this, the poor little infant was lying before us on the ground on its back; it had one eye open, and the other shut; its face was unwashed, and a small piece of dirty cloth was tied round its middle: the man, as he was talking to me, placed the child in the little furrow, pushed its arms to its side with the spade, and covering it so barely with earth that part of the cloth was still visible, he walked away and left it. I took the spade, and was going to bury the poor child myself, when I recollected that as a stranger I should probably give offence, and I therefore walked towards the gate. I met the same man, with an assistant, carrying a tray, in which was the body of a very old man, followed by his son, who was about forty; the party were all quarrelling, and remained disputing for some minutes after they had brought the body to the edge of the trench. This trench was about seven feet broad, and had been dug from one wall of the burial-ground to the other: the corpses were buried across it by fours, one above another, and there was a moveable shutter which went perpendicularly across the trench, and was moved a step forwards as soon as the fourth body was interred. One body had already been interred; the son jumped down upon it, and while he was thus in the grave, standing upon one body and leaning against three, the two grave-diggers gave him his father, who was dressed in a long, coarse, white linen shirt. The grave was so narrow that the man had great difficulty in laying the body in it, but as soon as he had done so, he addressed the lifeless corpse of his father, and embraced it with a great deal of feeling: the situation of the father and son, although so very unusual, seemed at the moment anything but unnatural. In scrambling out of the grave, the man very nearly knocked a woman out of the tier of corpses at his back; and as soon as he was up, the two attendants with their spades threw earth down upon the face and the white dress of the old man, until both were covered with a very thin layer of earth: the two men then jumped down with heavy wooden rammers, and they really rammed the corpse in that sort of way that, had the man been alive, he would have been killed; and we then all walked away.

MODE OF TRAVELLING.

THERE are two ways of travelling across the Pampas, in a carriage, or on horseback. The carriages are without springs, either of wood or iron, but they are very ingeniously slung on hide-ropes, which make them quite easy enough. There are two sorts of carriages, a long vehicle on four wheels, like a van (with a door behind), which is drawn by four or six horses, and which can carry eight people; and a smaller carriage on two wheels, of about half the length, which is usually drawn by three horses.

When I first went across the Pampas, I purchased for my party a large carriage, and also an enormous, two-wheeled, covered cart, which carried about twenty-five hundred weight of miners' tools, &c. I engaged a capataz (head-man), and he hired for me a number of peons, who were to receive thirty or forty dollars each for driving the vehicles to Mendoza.

The day before we started, the capataz came to me for some money to purchase hides, in order to prepare the carriages in the usual way. The hides were soaked, and then cut into long strips, about three-quarters of an inch broad, and the pole, as also almost all the wood-work of the carriage, were firmly bound with the wet hide, which, when dry, shrunk into a band almost as hard as iron. The spokes of the wheels, and, very much to our astonishment, the fellies or the circumference of the wheels were similarly bound, so that they actually travelled on the hide. We all declared it would be cut before it got over the pavement of Buenos Aires, but it went perfectly sound for seven hundred miles, and was then only cut by some sharp granite rocks over which we were obliged to drive.

With respect to provisions, we were told (truly enough) that there is little to be had on the Pampas but beef and water; and a quantity of provisions, with cherry brandy, &c. &c., was collected by the party, some of whom, I believe, fancied that I was going to take them, not to El Dorado, but to " that undiscovered country from which no travel-

ler returns;" however, when we were ready to start,
one of them found out that the loaves and fishes,
the canteen, &c., were all left out (whether by
accident or design, it matters not), and they then
all cheerfully consented to " rough it," which is
really the only way to travel without vexation in
any country. We took some brandy and tea with
us, but so destitute were we of other luxuries, that
the first day we had nothing to drink our tea out
of but egg-shells.

As it had been reported to the government of
Buenos-Ayres, that the Pampa Indians had invaded the country through which we had to pass,
the minister was kind enough to give me an order
to a Commandant who was on the road with troops,
for assistance if required; and besides this, we
purchased a dozen muskets, some pistols, and
sabres, which were slung to the roof of the carriage.

As it is customary to pay the peons half their
money in advance, and as men who have been paid
in advance have in all countries a number of
thirsty friends, it is very difficult to collect all the
drivers. Ours were of all colours, black, white,

and red; and they were as wild a looking crew as ever was assembled. We had six horses in the carriage, six in the cart, each of which was ridden by a peon, and I, with one of the party, rode.

The travelling across the Pampas a distance of more than nine hundred miles is really a very astonishing effort. The country, as before described, is flat, with no road but a track, which is constantly changed. The huts, which are termed posts, are at different distances, but upon an average, about twenty miles from each other; and in travelling with carriages, it is necessary to send a man on before, to request the Gauchos to collect their horses.

The manner in which the peons drive is quite extraordinary. The country being in a complete state of nature, is intersected with streams, rivulets, and even rivers, with pontanas (marshes), &c., through which it is absolutely necessary to drive. In one instance the carriage, strange as it may seem, goes through a lake, which of course is not deep. The banks of the rivulets are often very precipitous, and I constantly remarked that we drove over and through places which in Europe any military officer would, I believe, without hesitation report as impassable.

The mode in which the horses are harnessed is admirably adapted to this sort of rough driving. They draw by the saddle instead of the collar, and having only one trace instead of two, they are able, on rough ground, to take advantage of every firm spot; where the ground will only once bear, every peon takes his own path, and the horses' limbs are all free and unconstrained.

In order to harness or unharness, the peons have only to hook and unhook the lasso which is fixed to their saddle; and this is so simple and easy, that we constantly observed when the carriage stopped, that before any one of us could jump out of it, the peons had unhooked, and were out of our sight to catch fresh horses in the corral.

In a gallop, if any thing was dropped by one of the peons, he would unhook, gallop back, and overtake the carriage without its stopping for him. I often thought how admirably in practice this mode of driving would suit the particular duties of that noble branch of our army, the Horse Artillery.

The rate at which the horses travel (if there are

enough of them) is quite surprising. Our cart, although laden with twenty-five hundred weight of tools, kept up with the carriage at a hand-gallop. Very often, as the two vehicles were going at this pace, some of the peons, who were always in high spirits, would scream out, " Ah mi patron !" and then all shriek and gallop with the carriage after me; and very frequently I was unable to ride away from them.

But strange as the account of this sort of driving may sound, the secret would be discovered by any one who could see the horses arrive. In England, horses are never seen in such a state; the spurs, heels, and legs of the peons are literally bathed with blood, and from the sides of the horses the blood is constantly flowing rather than dropping.

After this description, in justice to myself, I must say, that it is impossible to prevent it. The horses cannot trot, and it is impossible to draw the line between cantering and galloping, or, in merely passing through the country, to alter the system of riding, which all over the Pampas is cruel.

The peons are capital horsemen, and several times we saw them at a gallop throw the rein on

48   MODE OF TRAVELLING.

the horse's neck, take from one pocket a bag of
loose tobacco, and with a piece of paper, or a leaf
of the Indian corn, make a segar, and then take
out a flint and steel and light it.

The post-huts are from twelve to thirty-six miles,
and in one instance fifty-four miles, from each other;
and as it would be impossible to drag a carriage
these distances at a gallop, relays of horses are sent
on with the carriage, and are sometimes changed
five times in a stage.

It is scarcely possible to conceive a wilder sight
than our carriage and covered cart, as I often saw
them *, galloping over the trackless plain, and preceded or followed by a troop of from thirty to
seventy wild horses, all loose and galloping, driven
by a Gaucho and his son, and sometimes by a couple of children. The picture seems to correspond

* I was one day observing them, instead of looking before
me, when my horse fell in a biscachero, and rolled over upon
my arm. It was so crushed that it made me very faint; but
before I could get into my saddle, the carriages were almost out
of sight, and while the sky was still looking green from the
pain I was enduring, I was obliged to ride after them, and I believe I had seven miles to gallop as hard as my horse could go,
before I could overtake the carriage to give up my horse.

with the danger which positively exists in passing through uninhabited regions, which are so often invaded by the merciless Indians.

*   *   *   *   *

In riding across the Pampas, it is generally the custom to take an attendant, and people often wait to accompany some carriage ; or else, if they are in condition, ride with the courier, who gets to Mendoza in twelve or thirteen days. In case travellers wish to carry a bed and two small portmanteaus, they are placed upon one horse, which is either driven on before, or, by a halter, tied to the postilion's saddle.

The most independent way of travelling is without baggage, and without an attendant. In this case, the traveller starts from Buenos Aires or Mendoza with a postilion, who is changed at every post. He has to saddle his own horses, and to sleep at night upon the ground on his saddle ; and as he is unable to carry any provisions, he must throw himself completely on the feeble resources of the country, and live on little else than beef and water.

It is of course a hard life; but it is so delightfully independent, and if one is in good riding condition, so rapid a mode of travelling, that I twice chose it, and would always prefer it ; but I recommend no one to attempt it, unless he is in good health and condition.

When I first crossed the Pampas, I went with a carriage, and although I had been accustomed to riding all my life, I could not at all ride with the peons, and after galloping five or six hours was obliged to get into the carriage; but after I had been riding for three or four months, and had lived upon beef and water, I found myself in a condition which I can only describe by saying that I felt no exertion could kill me. Although I constantly arrived so completely exhausted that I could not speak, yet a few hours' sleep upon my saddle, on the ground, always so completely restored me, that for a week I could daily be upon my horse before sunrise, could ride till two or three hours after sunset, and have really tired ten and twelve horses a day. This will explain the immense distances which people in South America are said to ride, which I am confident could only be done on beef and water.

At first, the constant galloping confuses the head,

and I have often been so giddy when I dismounted
that I could scarcely stand; but the system, by degrees, gets accustomed to it, and it then becomes
the most delightful life which one can possibly
enjoy. It is delightful from its variety, and
from the natural mode of reflecting which it en-
courages—for, in the grey of the morning, while
the air is still frosty and fresh, while the cattle are
looking wild and scared, and while the whole face
of Nature has the appearance of youth and innocence, one indulges in those feelings and speculations in which, right or wrong, it is so agreeable to
err; but the heat of the day, and the fatigue of
the body, gradually bring the mind to reason ;
before the sun has set many opinions are corrected,
and, as in the evening of life, one looks back with
calm regret upon the past follies of the morning.

In riding across the Pampas with a constant
succession of Gauchos, I often observed that the
children and the old men rode quicker than the
young men. The children have no judgment, but
they are so light, and always in such high spirits,
that they skim over the ground very quickly. The
old grey-headed Gaucho is a good horseman, with

great judgment, and although his pace is not quite
so rapid as the children's, yet, from being constant
and uniform, he arrives at his goal nearly in the
same time. In riding with the young men, I
found that the pace was unavoidably influenced by
their passions, and by the subject on which we
happened to converse; and when we got to the
post, I constantly found that, somehow or other,
time had been lost.

In crossing the Pampas it is absolutely necessary
to be armed, as there are many robbers or salteadores, particularly in the desolate province of
Santa Fé.

The object of these people is of course money,
and I therefore always rode so badly dressed, and
so well armed, that although I once passed through
them with no one but a child as a postilion, they
thought it not worth their while to attack me.
I always carried two brace of detonating pistols in
a belt, and a short detonating double-barrelled gun
in my hand. I made it a rule never to be an
instant without my arms, and to cock both barrels
of my gun whenever I met any Gauchos.

With respect to the Indians, a person riding can

use no precaution, but must just run the gauntlet, and take his chance, which, if calculated, is a good one.

If he fall in with them, he may be tortured and killed, but it is very improbable that he should happen to find them on the road; however, they are so cunning, and ride so quick, and the country is so uninhabited, that it is impossible to gain any information about them: besides this, the people are so alarmed, and there are so many constant reports concerning them, that it becomes useless to attend to any, and I believe it is just as safe to ride towards the spot at which one hears they are, as to turn back.

The greatest danger in riding alone across the Pampas, is the constant falls which the horses get in the holes of the biscachos.   I calculated, that, upon an average, my horse fell with me in a gallop once in every three hundred miles; and although, from the ground being very soft, I was never seriously hurt, yet previous to starting one cannot help feeling what a forlorn situation it would be, to break a limb, or dislocate a joint, so many hundred miles from any sort of assistance.

TOWN OF SAN LUIS.

* * * * *

FIFTH day (from Buenos Aires). We arrived
an hour after sunset—fortified post—scrambling in
the dark for the kitchen—cook unwilling—correo
(the courier) gave us his dinner—huts of wild-
looking people—three women and girls almost
naked*—their strange appearance as they cooked our
fowls. Our hut—old man immoveable—Maria or
Marequita's figure—little mongrel boy—three or
four other persons. Roof supported in the centre
by a crooked pole—holes in roof and walls—walls
of mud, cracked and rent—a water-jug in the corner
on a three-pronged stick—Floor, the earth—the
eight hungry peons, by moonlight, standing with
their knives in their hands over a sheep they were
going to kill, and looking on their prey like relentless tigers.

* " They be so wild as the donkey," said one of the Cornish
party, smiling; he then very gravely added, " and there be one
thing, sir, that I do observe, which is, that the farther we do go,
the wilder things do get !"

In the morning, morales and the peons standing by the fire—the blaze making the scene behind them dark and obscure—the horizon like the sea, except here and there the back of a cow to be seen —waggon and coach just discernable.

In the hut all our party occupied with the baggage—lighted by a candle crooked and thin— Scene of urging the patron (Master) to get horses, and Marequita to get milk—the patron wakening the black boy.

* * * * *

Twelfth day.—Left the post hut with three changes of horses to get to San Luis, distant thirty-six miles—inquired the way of one of the Gauenos who was drawing the carriage—he dismounted and traced it with his finger on the road—we were to turn off, when about three leagues, at a dead horse which we should see. I then galloped on with one of my party, knowing that we were to see no habitation until we got to San Luis—we had three hours and a half of day-light. About half way we began to think we had lost our path; however, we were sure to be wrong if we stopped to debate, and we therefore galloped on. Our horses got tired,

and the sun was nearly setting without any appearance of houses, but as the lower edge touched the
horizon, we discovered a hut, and riding up to it,
we were informed by a little girl that we were near
San Luis.   We got to the post just as it was dark,
and eagerly inquired of the wild group if there was
an inn in the town.   " No hai ! Senor ; no hai !"
We then inquired for beds.   " No hai ! Senor ; no
hai !"—" Is there a café ?"   " No hai ! Senor," in
exactly the same tone of voice.   When we looked
round us we found nothing but bare walls and fleas.
We happened (that day) to have English saddles,
and we therefore began to ask again about beds.
The woman told us we should have hers, and in a
few moments she brought mattress and all rolled
up, and laid it down on the floor ; however, when
I cast my eyes on the blanket, and above all the
sheets, I begged in the most earnest manner, that
she would let me have something a little cleaner.
" Son limpias," (they are clean) said the woman,
taking up the sheet, and pointing to a little spot
which looked whiter than the rest.   There was no
use in arguing the point, so I walked out of the
hut, leaving the corner of the sheet in the woman's

hand, and declaring that it was quite impossible to sleep there.

I went to the door of the Maestro de Posta (Postmaster), and told him that I had ridden all day without eating ; that I was very hungry, and begged to know what we could have: " Lo que quiere, Senor, tenemos todo," (whatever you choose, we have everything).

I knew too well what " todo" meant, and he accordingly explained to me that he had " carne de vacca and gallinas" (beef and fowls). I ordered a fowl, and then went to my room. The sight of the bed again haunted me, and after looking at it for some time with every inclination to persuade my-self that it was even bearable, but in vain—I resolved to go the Governor, deliver my letters, and see what I could do with him.

I procured a guide, who was to lead me in the dark to the Governor's house. After walking some distance, "Aqui sta," said the man. " What is that it?" said I, pointing to a door at which some black naked children were standing.—No, it was the next house.

The Governor was not at home, but I found his

wife sitting on a bed, surrounded by ladies—requested to sit down, but hurried off to the Coro-
nello—he was not at home, said a young lady, who
begged me to sit down—Went to the barracks—
my reception—an Ordenanza or soldier ordered to
return with me to the post, to desire the Postmaster
to treat me with particular respect—The town of
San Luis by moonlight—no houses to be seen, but
garden walls of mud—Went to look after my
dinner—found the girl who was to cook it sitting in
the smoke with the peons.—I saw a black iron pot
on the fire in which I supposed was my fowl—I
asked if the fowl was there? " No, Senor, aqui
sta," said the girl, throwing an old blanket off her
bare shoulders, and showing me the fowl alive in
her lap. I was going to complain, and I fear to
swear, but the smoke so got into my eyes and
mouth that I could neither see nor speak. At last
I asked for eggs, " No hai, Senor." " Good
heavens!" said I, " in the capital of San Luis is
there not one single egg?" " Yes," she said,
but it was too late, she would get me some mañana
(to-morrow). She asked me if I liked cheese.—
" Oh, yes," said I, eagerly.—She gave me an enor-

mous cheese, and insisted on my taking the whole of it, but she had no bread.

I had hurt my right arm by my horse falling; however, I carried the cheese into my room, and then did not know where to put it. The floor was filthy—the bed was worse, and there was nothing else; so supporting it with my lame arm, I stood for some seconds moralizing on the state of the capital of the Province of San Luis.

* * * * *

JOURNEY TO THE GOLD MINES AND LAVADEROS OF LA CAROLINA.

* * * * *

STARTED at day-break from San Luis, to go to the Gold Mines and Lavaderos* of La Carolina, which are in the mountains on the north of the town.

Drove a set of loose horses before us, and, about twelve o'clock, stopped to change.

The horses were driven to the edge of a precipice which was quite perpendicular, and which overhung a torrent, and we formed a semicircle about them while the peons began to catch them with the lasso, which they were much afraid of. The horses were so crowded and scared, that I expected they would all have been over the precipice: at last the hind-legs of one horse went down the cliff, and he hung in a most extraordinary man-

* Alluvial soil which is washed for gold.

ner by the fore-legs, with his nose resting on the
ground, as far from him as possible, to preserve
his balance. As soon as we saw him in this
situation, we allowed the other horses to escape,
and in a moment the peon threw his lasso with
the most surprising precision, and it went below the animal's tail like the breeching of harness. We then all hauled upon it, and at last
lifted the horse, and succeeded in dragging him
up : during the whole time time he remained
quiet, and to all appearance perfectly conscious
that the slightest struggle would have been fatal to
him. We then mounted our fresh horses, and
although the path over the mountains was so steep
and rugged, that we were occasionally obliged to
jump a foot or two from one level to another, we
scrambled along with the loose horses before us, at
the rate of nine or ten miles an hour.

In the evening, we came to a small stream of
water, which led us to the wretched hamlet of **La
Carolina,** which is close to the mine.

A man offered us a shed to sleep in, which we
readily accepted, and we then went into several of
the huts, and conversed with the poor people, who

had heard of rich English associations, and who
thought we were come to give them everything
they could desire.

In the evening we got some supper, and slept on
the ground in an out-house. We had observed a
very savage dog tied up in the yard, which was
constantly trying to get at us. In the middle of
the night, while the moon was shining upon us
through some holes in the roof, this dog walked
in, and after smelling us all, he went to sleep
among us.

The whole of the next day we spent in the mines
and the lavaderos, and in the evening I walked
alone into a little garden, and looked among the
soil for gold. I really was able to find a very few
particles, and it was singular to collect such a com-
modity in the gardens of such very poor people.

On my return I called at several of the huts, to
receive some gold dust which I had promised to
purchase of them. It happened that I had nothing
but a quantity of four-dollar gold-pieces, and although they were current all over South America,
I found, to my very great astonishment, that no one
here would take them. In vain I assured them of

their value, but these poor people (accustomed to change gold for silver) all shook their fingers in my face, and in different voices exclaimed " No vale nada," (gold is worth nothing,) and among such wild mountains, the great moral truth of their assertion rushed very forcibly into my mind.

I offered them the piece of four dollars for what they only asked two and three dollars, but they would not take it ; and we had hardly silver enough among us to remunerate our landlord for the board and lodging which he had afforded us.

Our horses which we had brought from San Louis were caught, and put into the corral the evening before we left the town, and they had consequently nothing to eat all that night.

The following day, as I have stated, we rode them sixty miles, and as it was then too late to turn them out, they were kept by the peon in the yard all that night.

The next day while we were inspecting the mines, they were turned out for four or five hours to graze among stones and rocks, where there was apparently nothing for them to eat, and they were then brought into the yard, where they remained

fasting all night. The next morning before day-
break we mounted them, and rode sixty miles back
to San Luis, and as some of the party came in
very late, I rather believe that the post-master
kept them in his corral all night, and that the following morning they were driven to the plain.

The poor creatures must of course have suffered
very much, but I did not know that at Carolina
there would have been nothing for them to eat; and
when we were there, I believe it was merciful to
them not to stay; but the truth is, that the business I was on was of such importance, that I really
had not time to think about it.

MENDOZA.

THE town of Mendoza is situated at the foot of the Andes, and the country around it is irrigated by cuts from the Rio de Mendoza. This river bounds the west side of the town, and from it, on the east side, there is a cut or canal about six feet wide, containing nearly as much water as would turn a large mill. This stream supplies the town with water, and at the same time adorns and refreshes the Almeida or public walk. It waters the streets which descend from it to the river, and can also be conducted into those which are at right angles.

Mendoza is a neat small town, built upon the usual plan. The streets are all at right angles; there is a plaza or square, on one side of which is a large church, and several other churches and convents are scattered over the town. The houses are only one story high, and all the principal ones have a porte cochére, which enters a small

F

court, round the four sides of which the house
extends.

The houses are built of mud, and are roofed
with the same. The walls are white-washed, which
gives them a neat appearance, but the insides of
the houses, until they are white-washed, look like
an English barn. The walls are of course very
soft; occasionally a large piece of them comes off,
and they are of that consistency, that, in a very
few moments, a person, either with a spade or a
pick-axe, could cut his way through any wall in
the town. Several of the principal houses have
glass in the window-sashes, but the greatest number have not. The houses are almost all little
shops, and the goods displayed are principally
English cottons.

The inhabitants are apparently a. very quiet,
respectable set of people. The Governor, who is
an old man, has the manners and the appearance of
a gentleman: he has a large family of daughters,
who are pleasing-looking girls. The men are
dressed in blue or white jackets, without skirts.
The women are only seen in the day sitting at their
windows, in complete dishabille, but in the evening

they come upon the Almeida, dressed with a great
deal of taste, in evening dresses and low gowns,
and completely in the costume of London or Paris.
The manner in which all the people seem to asso-
ciate together, shews a great deal of good feeling
and fellowship, and I certainly never saw less apparent jealousy in any place.

The people, however, are extremely indolent. A
little after eleven o'clock in the morning, the shop-
keepers make preparations for the siesta; they
begin to yawn a little, and slowly to put back the
articles which they have, during the morning, displayed on their tables. About a quarter before
twelve they shut up the shops, the window-shutters
throughout the town are closed, or nearly so, and
no individual is to be seen until five, and sometimes
until six o'clock, in the evening.

During this time I used generally to walk about
the town to make a few observations. It was really
singular to stand at the corner of the right-angled
streets, and in every direction to find such perfect
solitude in the middle of the capital of a province.
The noise occasioned by walking was like the echo
which is heard in pacing by oneself up the long

aisle of a church or cathedral, and the scene resembled the deserted streets of Pompeii.

In passing some of the houses I often heard
people snoring, and when the siesta was over, I
was often much amused at seeing the people awaken,
for there is infinitely more truth and pleasure in
thus looking behind the scenes of private life, than
in making formal observations on man when dressed
and prepared for his public performance. The
people generally lie on the ground or floor of the
room, and the group is often amusing.

I saw one day an old man (who was one of the
principal people in the town) fast asleep and happy.
The old woman his wife was awake, and was sitting
up in easy dishabille scratching herself, while her
daughter, who was a very pretty-looking girl of
about seventeen, was also awake, but was lying on
her side kissing a cat.

In the evening the scene begins to revive. The
shops are opened; a number of loads of grass are
seen walking about the streets, for the horse that is
carrying them is completely hid. Behind the load
a boy stands on the extremity of the back; and to
mount and dismount he climbs up by the animal's

tail. A few Gauchos are riding about, selling fruit ; and a beggar on horseback is occasionally seen, with his hat in his hand, singing a psalm in a melancholy tone.

As soon as the sun has set, the Almeida is crowded with people, and the scene is very singular and interesting. The men are sitting at tables, either smoking segars or eating ices : the ladies are sitting on the mud benches which are on both sides of the Almeida. This Almeida is a walk nearly a mile long, between two rows of tall poplars ; on one side of it are the garden-walls of the town, concealed by roses and shrubs, and on the other the stream of water which supplies the town.

It will hardly be credited that, while this Almeida is filled with people, women of all ages, without clothes of any sort or kind, are bathing in great numbers in the stream which literally bounds the promenade. Shakespeare tells us, that " the chariest maid is prodigal enough if she unveil her beauties to the moon," but the ladies of Mendoza, not contented with this, appear even before the sun ; and in the mornings and evenings they really bathe without any clothes in the Rio de Mendoza,

the water of which is seldom up to their knees, the
men and women all together; and certainly, of all
the scenes which in my life I have witnessed, I
never beheld one so indescribable.

However, to return to the Almeida :—the walk
is often illuminated in a very simple manner by
paper lamps, which are cut into the shapes of stars,
and are lighted by a single candle. There is generally a band of music playing, and at the end of
the walk is a temple built of mud, which is very
elegant in its form, and of which it may truly be
said " materiam superabat opus."

The few evenings I was at Mendoza, I always
went as a complete stranger to this Almeida to eat
ices, which, after the heat of the day, were exceedingly delightful and refreshing; and as I put spoonful after spoonful into my mouth, looking above
me at the dark outline of the Cordillera, and listening to the thunder which I could sometimes hear
rumbling along the bottoms of the ravines, and
sometimes resounding from the tops of the mountains, I used always to acknowledge, that if a man
could but bear an indolent life, there can be no
spot on earth where he might be more indolent and

more independent than at Mendoza, for he might sleep all day, and eat ices in the evening, until his hour-glass was out. Provisions are cheap, and the people who bring them quiet and civil; the climate is exhausting, and the whole population indolent— " Mais que voulez-vous ? " how can the people of Mendoza be otherwise ? Their situation dooms them to inactivity ;—they are bounded by the Andes and by the Pampas, and, with such formidable and relentless barriers around them, what have they to do with the history, or the improvements, or the notions of the rest of the world ? Their wants are few, and nature readily supplies them,— the day is long, and therefore as soon as they have had their breakfasts, and have made a few arrangements for their supper, it is so very hot that they go to sleep, and what else could they do better ?

THE PAMPAS.

RETURNED to the Fonda in the evening at ten
o'clock, and found the two horses standing in the
yard with nothing to eat, and a young Gaucho,
who was to accompany me as postilion, lying on the
ground asleep on his saddle. Next morning before
day-break, got up, saddled my horse, and with my
saddle as my bed, and some pistols and money,
commenced my gallop for Buenos Ayres.

Country to be described—delightful feeling of
independence at the mode of travelling—air frosty,
and ground hard.—The sun rose, and shortly after
got to the first post.—Had a letter for the lady
from her husband who was at Mendoza—went to
give it to her, while the Gaucho who was to accompany me was driving the horses into the corral—
found the lady in bed.—" Siente se, Senor," said
she, pointing to an old chair which was at the head
of the bed—sat down, and told her the letter was

from her husband—she placed it under her pillow, and then offered me some maté, but I had no time to wait for it, and started.

At third post from Mendoza, the post-master, who might be exhibited in England as a curious specimen of an indolent man, to every thing I said, replied " *si* "—it was but an aspiration, and he seemed never to have said any other word—I had twice passed his house, and it was always the same *Si !*

Galloped on with no stopping, but merely to change horses until five o'clock in the evening— very tired indeed, but on coming to the post-hut, saw the horses in the corral, and resolved to push on.—Started with a fresh horse, and a young Gaucho, who, singing as he went, galloped like the wind ; the sun set, and it got so dark, that, for more than an hour, I expected that every moment the boy would get away from me, as the road was rough, and through wood. At half-past seven, after having galloped a hundred and fifty-three miles, and been fourteen hours and a half on horse-back, got to the post—found the hut occupied by some people who had arrived in a carriage—quite

exhausted—nothing to eat—asked for bread, they
had none—I really could scarcely speak—carried
my saddle into a shed—two children asleep, and
one black girl—lay down upon the ground, and
instantly fell asleep—was awakened in two or three
hours by the woman of the post, who had brought
me some soup with meat in it—eat it all up, and
again dropt off to sleep—an hour before daylight
was awakened by the Gaucho who was to go with
me. "Vamos, Senor!" said he, in a sharp, impa-
tient tone of voice—got up, had some maté,
mounted my horse, and as I galloped along felt
pleased that the sun which had left me the evening
before thirty miles nearer Mendoza, should find me
at my work. At first post detained fifteen minutes
for horses—the stage the longest between Mendoza
and Buenos Ayres, being fifty-one miles—the
woman would only give me one spare horse, which
we drove before us. Galloped my horse till he came
to a stand-still, and then got on the fresh one, and
left the postilion behind. In about an hour this other
horse quite done up—by constant spurring could
just keep him in a canter—at last down he fell,
and my foot hung in the stirrup—my long spur

was also entangled in the sheep-skin which was above my saddle—saw by the panting of the horse's flank and nostrils that he was too tired to be off with me. Mounted and cantered him till he fell down on my other leg, and I was then lame on both legs—overtook a boy driving some loose horses—took one of them, and my horse was driven among the flock, until we came to the post. Postmaster very kind, and ordered a Gaucho to give me an easy-going horse, as both my legs hurt me very much—started with a boy, but our horses were done up before we got to San Luis—obliged to walk part of the distance, and then by kicking and spurring got into San Luis just as the sun set.—See description of the post-house and town of San Luis.

At San Luis was advised by groups of people, not to go on by myself, as the courier and postilion (from Buenos Aires), with their horses and a dog, had just been found on the road with their throats cut—advised to join the courier who was just setting out for Buenos Aires. Accordingly, next morning started with the courier and three peons as guards, all armed with old pistols

and guns. Courier a little old man of about fifty-
five years of age—had been riding all his life—had
a face like a withered apple—carried his pistol in
his hand—told me he was father to the courier who
had just been murdered—that he was his only son
—that he had just succeeded in getting him the
appointment—that he was nineteen—and that it was
his first journey as courier—that he had no pistols,
not even a knife—that it was barbarous to kill him
—that he must have died like a lamb, &c. &c.
This story he repeated at every post-hut, and
people were so fond of asking for it, and he so
willing to give it, that we lost many minutes at
each post. He would relate it to anybody—at
one post he told it to a great rough mongrel-look-
ing fellow, who was sitting on a stone while a little
girl was combing his woolly hair—" *En dos?*" said
the little girl who had divided his hair at the back
of his head, and who proposed to plait it into two
tails—" Si!" grunted her father, half asleep, and
nodding his head, as he listened to the courier's
story. We therefore rode all day, and only went a
hundred and two miles.—Next morning off before
sunrise, and took a postilion, and travelling by

myself got on much quicker, but the horses still weak, and in the whole day could only proceed a hundred and ten miles.

Two more days rode from morning till night, sleeping on the ground, with nothing to eat but beef—at last came to that part of the province of Santa Fé near which the courier had been murdered. The post-master refused to give me horses to go on unless I could find a guard, as he said the postilions would not go by themselves; he insisted on my waiting for the courier, and I accordingly lost half a day, as he did not arrive till night. Next morning at day-break got up—saw the poor old courier lying on his saddle—he had a segar in his mouth, and for a long time he remained on his back praying and crossing himself—Started with the master of the post, an additional Gaucho, and the postilion, all armed—very little conversation. As we approached the spot, it appeared as if they all expected that the Salteadores (robbers) would be there—after riding some leagues, left the road, and galloped through the dry grass towards a small black-looking hut in ruins. It was one of those which had been burnt by the Indians, and the whole

family had been murdered in it. When we got to
it, I looked around me, and no other hut or habitation was to be seen; there were no cattle, and when
a few *gamas* (deer), which for a few moments were
in sight, had fled away, we were left completely to
ourselves, and not a bird or any animal was to be
seen. We were in the centre of a deserted province. We galloped up to the hut—it was built of
large unbaked bricks and mud : the roof had been
burnt—one of the gables had fallen to half its
height—the other looked nearly falling—one wall
had fallen, and we all rode up to this side of the
hut—Close to us there was a deep well, into which
the Salteadores had thrown all the bodies—first the
courier and postilion, then the dog, and then the
horses. The carcases of the horses lay before us—
they were nearly eaten up by the eagles and biscachos. The dog had not been touched—he was a
very large one—and from the heat of the weather,
he was now bloated up to a size quite extraordinary
—his throat was cut, and in my life I never saw so
much expression in the countenance of a dead
animal—his lip was curled up, and one could not
but fancy that it expressed the feelings of rage and

fidelity under which he had evidently fought to the last. In the hut lay the bodies of the courier and postilion with their throats cut *—they were barely covered over with some of the loose bricks from the wall. Some pieces of the courier's poncho were lying about, as also several of the covers of the letters which the murderers had opened. In the centre of the hut were the white ashes of a fire which they had kindled—at the corner of the hut tood a solitary peach-tree in blossom—its contrast with the scene before us was very striking. The old courier said something to the post-master, who immediately climbed upon the ruined wall, and threw down some loose bricks—he fell—burst of laughter—we all got off our horses, and we covered the bodies over with bricks—" Con que, Senores," said the old man, " haremos un oracion para el defunto"—we all took off our hats, and stood round the pile—opposite were our horses looking at us—the old man had thrown the handkerchief off his head, and his beard, which was of four days growth, was quite white—he stood over the body

* They had been taken out of the well by some Gauchos.

of his only son, and offered up some prayer, to
which all the Gauchos joined their responses. I
joined and crossed myself with them, for as the
courier looked at me, I was anxious to assist in
alleviating the sorrows of an old man, and entertaining my own feelings, which it is not necessary
to describe.

As soon as the ceremony was over (it lasted about
two minutes), we put on our hats. " Con que,
Senores," said the old man ; and after a long pause,
" Vamos!" said he, upon which the party split
into groups to light segars. I had scarcely lighted
mine, when the old man came up to light his. His
son's body was at our feet, but he put his face
close to mine, and as he was sucking and blowing,
with that earnestness of countenance which is only
known to those who are [in the habit of lighting a
segar, I could not help thinking what an odd
scene was before me. However we mounted our
horses—I took a last farewell look at the peach-
tree, and we then all galloped across the dry
brown grass, to regain the road, and the few minutes of time which we had thus spent at the hut]

*   *   *   *   *

* * * * *

At some distance I saw a boy riding through the thistles and clover, and as he was swinging horizontally above his head the bolas or balls, I perceived he was hunting for ostriches, and I therefore rode up to him.

He was a black boy of about fourteen years of age, slight, and well-made, but with scarcely anything on except the remains of a scarlet poncho. I asked him several questions—where he expected to find the ostriches, &c. &c. &c., to which he gave me no answer, but continued swinging the balls round his head, and looking about him. I was asking him some other insignificant questions, when he cut me short, by asking me if I would sell my spurs; and before I had time to reply, an ostrich was in sight, and he darted away from me like an arrow. I was, just at the moment, among a group of biscacheros—my horse fell, and before I had got clear of them, the boy was on the horizon, and from the contempt with which he had left me, I did not feel inclined to follow him.

* * * * *

*   *   *   *   *

The biscacho is found all over the plains of the
Pampas. Like rabbits, they live in holes which
are in groups in every direction, and which make
galloping over these plains very dangerous.   The
manner, however, in which the horses recover
themselves, when the ground over these subterra-
nean galleries gives way, is quite extraordinary.
In galloping after the ostriches, my horse has
constantly broken in, sometimes with a hind leg,
and sometimes with a fore one ; he has even come
down on his nose, and yet recovered : however, the
Gauchos occasionally meet with very serious acci-
dents.   I have often wondered how the wild horses
could gallop about as they do in the dark, but I
really believe they avoid the holes by smelling
them, for in riding across the country, when it has
been so dark that I positively could not see my
horse's ears, I have constantly felt him, in his gallop,
start a foot or two to the right or left, as if he had
trod upon a serpent, which, I conceive, was to avoid
one of these holes.   Yet the horses do very often
fall, and certainly, in the few months I was in the

Pampas, I got more falls than I ever before had, though in the habit of riding all my life. The Gauchos are occasionally killed by these biscache-ros, and often break a limb.

In the middle of the Pampas I once found a Gaucho standing at the hut, with his left hand resting on the palm of his other hand, and appa-rently suffering great pain. He told me his horse had just fallen with him in a biscachero, and he begged me to look at his hand. The large muscle of the thumb was very much swelled, and every time I touched it with my fore-finger, the poor fellow opened his mouth, and lifted up one of his legs. Being quite puzzled with one side of his hand, I thought I would turn it round, and look at the other side, and upon doing so, it was in-stantly evident that the thumb was out of joint. I asked him if there was any doctor near ; the Gau-cho said he believed there was one at Cordova, but as it was five hundred miles off, he might as well have pointed to the moon. " Is there no person," said I, " nearer than Cordova, that understands anything about it ? " " No hai, Senor," said the poor fellow. I asked him what he intended to do

with his thumb: he replied that he had washed it
with salt and water, and then he earnestly asked
me if that was good for it?   " Si! si! si!" said I,
walking away in despair, for I thought it was useless to hint to him, that " not all the water in the
wide rude sea" would put his thumb into its joint;
and although I knew it ought to be pulled, yet one
is so ignorant of such operations, that not knowing
in what direction, I therefore left the poor fellow
looking at his thumb, in the same attitude in which
I found him   But, to return to the biscachos.

These animals are never to be seen in the day,
but as soon as the lower limb of the sun reaches
the horizon, they are seen issuing from their holes
in all directions, which are scattered in groups like
little villages all over the Pampas.   The biscachos,
when full grown, are nearly as large as badgers;
but their head resembles a rabbit, excepting that
they have very large bushy whiskers.

In the evening they sit outside their holes,
and they all appear to be moralising.   They are
the most serious-looking animals I ever saw, and
even the young ones are grey-headed, have mus-
tachios, and look thoughtful and grave.

In the day-time their holes are always guarded by two little owls, who are never an instant away from their post. As one gallops by these owls, they always stand looking at the stranger, and then at each other, moving their old-fashioned heads in a manner which is quite ridiculous, until one rushes by them, when fear gets the better of their dignified looks, and they both run into the biscacho's hole.

*　　*　　*　　*　　*

THE PAMPAS—PROVINCE OF SANTA FE'.

TRAVELLING from Buenos Aires to Mendoza by myself, with a *virloche*, or two-wheeled carriage— entrance behind—two side-seats—had two peons— Pizarro, who had already travelled twelve hundred miles, and Cruz, a friend of Pizarro, had travelled for three days a hundred and twenty miles a day— Pizarro's fidelity and attention—at night when he got in, his dark black face tired, and covered with dust and perspiration—his tongue looked dry, and his whole countenance jaded—yet his frame was hard as iron. His first object at night to get me something to eat—to send out for a live sheep—He made a fire and cooked my supper—as soon as I had supped, he brought me a candle at the carriage door, and watched me while I undrest to sleep there—then wished me good night, got his own supper, and slept on his saddle at the wheel of the carriage. As soon as I awoke, and, before day-light, anxious to get on, I used to call out " Pi-

zarro!" "Aqui sta l'agua, Senor," said he, in a
patient low tone of voice—he knew I liked to have
water to wash in the morning, and he used to get
it for me, sometimes in a saucer, sometimes literally
in a little maté cup, which did not hold more than
an egg-shell, and in spite of his fatigue he was
always up before I awoke, and waiting at the door
of the carriage till I should call for him.

Province of Santa Fé to be described—its wild,
desolate appearance—has been so constantly ra-
vaged by the Pampas Indians, that there are now
no cattle in the whole province, and people are
afraid to live there. On the right and left of the
road, and distant thirty and forty miles, one occasionally sees the remains of a little hut which has
been burnt by the Indians, and as one gallops
along, the Gaucho relates how many people were
murdered in each—how many infants slaughtered
—and whether the women were killed or carried
away. The old post-huts are also burnt—new ones
have been built by the side of the ruins, but the
rough plan of their construction shews the insecu-
rity of their tenure. These huts are occupied only
by men, who are themselves generally robbers, but

in a few instances their families are living with them.
When one thinks of the dreadful fate which has
befallen so many poor families in this province,
and that any moment may bring the Indians again
among them, it is really shocking to see women
living in such a dreadful situation—to fancy that
they should be so blind, and so heedless of experience;—and it is distressing to see a number of innocent little children playing about the door of a hut,
in which they may be all massacred, unconscious of
the fate that may await them, or of the blood-thirsty,
vindictive passions of man.

We were in the centre of this dreary country—I
always rode for a few stages in the morning, and I
was with a young Gaucho of about fifteen years of
age, who had been born in the province—his father
and mother had been murdered by the Indians—he
had been saved by a man who had galloped away
with him, but he was then an infant, and remembered nothing of it. We passed the ruins of a hut
which he said had belonged to his aunt—he said
that, about two years ago, he was at that hut with his
aunt and three of his cousins, who were young men
—that while they were conversing together a boy gal-

loped by from the other post, and in passing the door
screamed out, " Los Indios! los Indios !"—that he
ran to the door, and saw them galloping towards the
hut without hats, all naked, armed with long lances,
striking their mouths with their bridle hands, and uttering a shriek, which he described as making the
earth tremble—he said that there were two horses
outside the hut, bridled, but not saddled—that he
leapt upon the back of one and galloped away—that
one of the young men jumped on the other, and
followed him about twenty yards, but that then he
said something about his mother, and rode back to
the hut—that just as he got there the Indians surrounded the hut, and that the last time he saw his
cousins they were standing at the door with their
knives in their hands—that several of the Indians
galloped after him, and followed him more than a
mile, but that he was upon a horse which was
" muy ligéro, (very swift) muy ligéro," said the
boy ; and as we galloped along he loosened his
rein, and darting on before me, smiled at shewing
me the manner in which he escaped, and then
curbing his horse to a hand-gallop, continued his
history.

He said that when the Indians found he was getting away from them, they turned back—that he escaped, and that when the Indians had left the province, which was two days after, he returned to the hut. He found it burnt, and saw his aunt's tongue sticking on one of the stakes of the corral; her body was in the hut; one of her feet was cut off at the ancle, and she had apparently bled to death. The three sons were outside the door naked, their bodies were covered with wounds, and their arms were gashed to the bone, by a series of cuts about an inch from each other, from the shoulder to the wrist.

The boy then left me at the next post, and I got into the carriage—the day getting hot, and the stage twenty-four miles. After galloping about an hour, I saw a large cloud of smoke on the horizon before me; and as the Indians often burn the grass when they enter the country, I asked Pizarro what it was? He replied, " Quien sabe,—Senor, what it may be;" however, on we galloped.

I took little notice of it, and began to think of the dreadful story the boy had told me, and of many similar ones which I had heard; for I had

always endeavoured to get at the history of the
huts which were burnt, although I always found
that the Gauchos thought very little about it ; and
that the story was sometimes altogether in oblivion,
before time had crumbled into dust the tottering
mud walls which were the monuments of such
dreadful cruelties.

It appears that the Pampas Indians, who, in
spite of their ferocity, are a very brave and handsome race of men, occasionally invade " los Cris-
tianos," as the Gauchos always term themselves,
for two objects—to steal cattle, and for the pleasure of murdering the people ; and that they will
even leave the cattle to massacre their enemies.

In invading the country, they generally ride all
night, and hide themselves on the ground during
the day ; or, if they do travel, crouch almost
under the bellies of their horses, who by this means
appear to be dismounted and at liberty. They
usually approach the huts at night at a full gallop, with their usual shriek, striking their mouths
with their hands—and this cry, which is to inti-
midate their enemies, is continued through the
whole of the dreadful operation.

Their first act is to set fire to the roof of the hut, and it is almost too dreadful to fancy what the feelings of a family must be, when, after having been alarmed by the barking of the dogs, which the Gauchos always keep in great numbers, they first hear the wild cry which announces their doom, and in an instant afterwards find that the roof is burning over their heads.

As soon as the family rush out, which they of course are obliged to do, the men are wounded by the Indians with their lances, which are eighteen feet long, and as soon as they fall they are stripped of their clothes; for the Indians, who are very desirous to get the clothes of the Christians, are careful not to have them spoiled by blood. While some torture the men, others attack the children, and will literally run the infants through the body with their lances, and raise them to die in the air. The women are also attacked, and it would form a true but a dreadful picture to describe their fate, as it is decided by the momentary gleam which the burning roof throws upon their countenances.

The old women, and the ugly young ones, are instantly butchered; but the young and beautiful are idols, by whom even the merciless hand of the savage is arrested. Whether the poor girls can ride or not, they are instantly placed upon horses, and when the hasty plunder of the hut is concluded, they are driven away from its smoking ruins, and from the horrid scene which surrounds it.

At a pace which in Europe is unknown, they gallop over the trackless regions before them, fed upon mare's flesh, sleeping on the ground, until they arrive in the Indians' territory, when they have instantly to adopt the wild life of their captors.

I was informed by a very intelligent French Officer, who was of high rank in the Peruvian army, that, on friendly terms, he had once passed through part of the territory of these Pampas Indians, in order to attack a tribe who were at war with them, and that he had met several of the young women who had been thus carried off by the Indians.

He told me that he had offered to obtain permission for them to return to their country, and that he had, in addition, offered them large sums of money if they would, in the mean while, act as interpreters ; but they all replied, that no induce-

ment in the world should ever make them leave
their husbands, or their children, and that they
were quite delighted with the life they led.

While I was sitting upon the side seat of the
carriage, reflecting on the cruelties which had been
exercised in a country which, in spite of its history,
was really wild and beautiful, and which possessed
an air of unrestrained freedom which is always ex-
hilarating, I remarked that the carriage was only
at a walk, an occurrence which in South America
had never before happened to me, and in an instant
it stopped. " Vea, Senor," said Pizarro, with a firm
countenance, as he turned back to speak to me,
" que tanta gente !" he pointed with his right hand
before him, and I saw that the smoke which I had
before observed was dust, and in it I indistinctly
saw a crowd of men on horseback in a sort of wild
military array ; and on both flanks, at a great distance off, individual horsemen, who were evidently
on the look out to prevent a surprise. Our horses
were completely tired; the whole body were
coming rapidly towards us, and to mend the matter,
Pizarro told me that he was afraid they were los
Indios. " Senor," said he, with great coolness, and

yet with a look of despair, " Tiene armas á fuégo ?"
I told him I had none to spare, for I had only a short
double-barrelled gun and two brace of pistols. "Aqui
un sable, Pizarro!" said I, pushing the handle of a
sabre towards him from the window of the carriage.
"Que sable !" said he, almost angrily; and raising his
right arm perpendicularly over his head, in a sort of
despair, he added, "contra tanta gente!" but while
his arm was in the position described, " Vamos !"
said he, in a tone of determined courage, and giving
his hand half a turn, he spurred his jaded horse,
and advanced instantly at a walk. Poor Cruz, the
other peon, seemed to view the subject altogether
in a different light; he said not a word, but as I
cast a glance at him, I perceived that his horse, far
from pulling the carriage, was now and then hanging back a little—a just picture of his rider's feelings. I could not help for a moment admiring
Pizzaro's figure, as I saw him occasionally digging
his spurs into the side of his horse, which had me,
the carriage, Cruz, and his horse to draw along;
however, I now began to think of my own situation.

I earnestly wished I had never come into the

country, and thought how unsatisfactory it was to be tortured and killed by mistake in other people's quarrels—however, this would not do. I looked towards the cloud of dust, and it was evidently much nearer. In despair, I got my gun and pistols, which were all loaded, and when I had disposed of them, I opened a small canvas bag which contained all my ammunition gimcracks, for my gun and pistols had all fulminating locks. I ranged all on the seat before me—the small powder-flask, the buck-shot, the bullets, the copper caps, and the punched cards; but the motion of the carriage danced them all together, and once or twice I felt inclined, in despair, to knock them all off the seat, for against so many people resistance was useless; however, on the other hand, mercy was hopeless, so I, at last, was driven to make the best of a very bad bargain.

The carriage, which had a window at each of the four sides, had wooden blinds, which moved horizontally. I therefore shut them all, leaving an embrasure of about two inches, and then for some seconds I sat looking at the crowd which was coming towards us.

As they came close to us, for until then I could scarcely see them for dust, I perceived that they had no spears, and next that they wore clothes; but as they had no uniforms I conceived that they were a crowd of Montaneros, who are quite as cruel as the Indians: however, as soon as they came to us, and when some of them had passed us, Pizarro pulled up and talked to them. They were a body of seven hundred wild Gauchos, collected and sent by the governors of Cordova and some other provinces, to proceed to Buenos Aires to join the army against the Brazilians; and on their flank they had scouts, to prevent a surprise by the Indians, who had invaded the country only a few weeks before.

It was really a reprieve; every thing I saw for the rest of the day pleased me—and for many days afterwards, I felt that I was enjoying a new lease of my life.

———

H

THE PAMPAS.

* * * * *

Two days afterwards, I was riding near the carriage, which was galloping along—Pizarro and Cruz looking fatigued and dirty, while the postilion before them, fresh and careless, was singing a Spanish song, when Pizarro's horse fell, and although Cruz tried to pull up, the postilion's horse dragged Pizarro along the ground at least twenty yards.

I really thought he was killed; however he quietly declared he was not hurt, and, without saying one other word, he adjusted his saddle, and galloped on to the next stage. As he was there mounting a young horse, which evidently had scarcely ever before been saddled, the creature plunged very violently. Pizarro was evidently weak from his accident, and, as he fell, the horse kicked him with both legs on his back.

Still he declared he was not hurt, though he looked very faint, and could scarcely mount his horse. I galloped on to the post-hut, and waited there more than an hour for the carriage. At last I saw it coming at a walk, and as soon as it drove up, Pizarro said he could go no farther. I was therefore obliged to order another boy as a postilion, and while they were catching the horses with the lasso, I was assisting poor Pizarro. I was very sorry to be obliged to leave him, particularly as he seemed so unwilling to leave me. I gave him some money, half a bottle of brandy, which was all I had ; and to a woman, who was a few years younger than Pizarro, and of the same mongrel breed as himself, I gave two dollars, to rub Pizarro's back three times a-day with the brandy; and I put some salt into it, that the woman should not drink the brandy, instead of rubbing Pizarro's back with it. This being all I could do for him, I mounted my horse, and wishing him good-bye, to which he replied, " Senor, vaya con Dios," I left him.

I desired the carriage to follow, and I rode from post to post ordering horses to be ready for the carriage, and got to San Luis about one o'clock in the morning. I was completely by myself, with-

out any postilion ; but, as it was a fine moon-light
night, and as I had twice before travelled over the
country, I managed to go the right road, and at
five o'clock I again started to ride towards Mendoza.

*   *   *   *   *

THE PAMPAS.

*   *   *   *   *

In the province of Santa Fé, a few of the posts
are fortified to protect the inhabitants against the
Indians.

The fort is a very simple one. The huts are
surrounded by a small ditch, which is sometimes
fenced on the inside with a row of prickly pears.

The ditch I have generally been able to jump
over on foot, but no horse of the country would
attempt to leap it.

Most of these forts have often been attacked by
the Indians; and one of them I looked at with peculiar interest, as it had very lately been defended
for nearly an hour by eight Gauchos against about
three hundred Indians. The cattle, the women,
and six families of little children, were all in the
inside, spectators of a contest on which so much
depended, and they described their feelings to me
with a great deal of nature and expression.

They said that the naked Indians rode up to the

ditch with a scream which was quite terrific, and that, finding that they could not cross it, the Cacique at last ordered them to get off their horses and pull down the gate. Two had dismounted, when the musket which the Gauchos had, and which before had constantly missed fire, went off, and one of the Indians was shot. They then all galloped away; but in a few seconds their Cacique led them on again with a terrible cry, and at a pace which was indescribable. They took up their dead comrade and then rode away, leaving two or three of their spears on the ground.

One of these long spears was leaning against the hut, and as the Gauchos who had defended the place stood by it, muffled up in their ponchos, with two or three women suckling their infants, several children playing about them, and three or four beautiful girls looking up to them, I thought the spear was one of the proudest military trophies I had ever beheld.

I could never learn that any of these forts had been taken by the Indians, who can do nothing on foot, and whose horses cannot leap; but the ditches are so shallow and so narrow, that by killing a few horses, and tumbling them in, they might in two minutes ride into any part of the place.

I often asked the Gauchos why they did not defend themselves in the corral, which at first appeared to me to be a stronger position than the forts; but they said that the Indians often brought lassos of hide, with which they could pull down the stakes of the corral; that sometimes they made a fire against them, and that, besides this, their spears being eighteen feet in length, they were often able to kill every animal in the corral.

*   *   *   *   *

THE PAMPAS.

*　　*　　*　　*　　*

THE fear which all wild animals in America have of man is very singularly seen in the Pampas. I often rode towards the ostriches and gamas, crouching under the opposite side of my horse's neck; but I always found that, although they would allow any loose horse to approach them, they, even when young, ran from me, though little of my figure was visible; and when one saw them all enjoying themselves in such full liberty, it was at first not pleasing to observe that one's appearance was everywhere a signal to them that they should fly from their enemy. Yet it is by this fear that "man hath dominion over the beasts of the field," and there is no animal in South America that does not acknowledge this instinctive feeling.

As a singular proof of the above, and of the difference between the wild beasts of America and of the Old World, I will venture to relate a circumstance

which a man sincerely assured me had happened to
him in South America.

He was trying to shoot some wild ducks, and, in
order to approach them unperceived, he put the
corner of his poncho (which is a sort of long narrow
blanket) over his head, and crawling along the
ground upon his hands and knees, the poncho not
only covered his body, but trailed along the ground
behind him. As he was thus creeping by a large
bush of reeds, he heard a loud sudden noise, between a bark and roar : he felt something heavy
strike his feet, and instantly jumping up, he saw,
to his astonishment, a large male lion actually
standing on his poncho, and perhaps the animal
was equally astonished to find himself in the immediate presence of so athletic a man !

The man told me he was unwilling to fire, as his
gun was loaded with very small shot, and he therefore stood his ground, and the lion stood on his
poncho for many seconds ; at last he turned his
head, and walking very slowly away about ten
yards, he stopped and turned again. The man
still stood his ground, upon which the lion tacitly
acknowledged his supremacy, and walked off.

THE PAMPAS.

* * * * *

AFTER being in the post-hut a few minutes, I heard a sigh, and looking into the corner from whence it proceeded, I saw an old sick woman lying on the ground. Her head was resting on a horse's skull, close to a great hole in the wall, and when she earnestly asked me if I had any thing " por remedio," I instantly advised her to move herself into a warmer corner. She was feverish and ill, and seemed disappointed at the advice I had given her—she did not understand what the hole in the wall could possibly have to do with her illness, and she again asked me if I had any " remedio."

I had in my waistcoat pocket a little dirty paper parcel of calomel and jalap, which I had promised, much against my will, to carry with me, and which I had already twice carried across the Pampas. I did not exactly know how much there was of it, but I had a great mind to shake a little of it into the old woman's mouth, for I thought (as she had certainly never tasted calomel before) it would pro-

bably work a miracle within her ; however, she
was so ill that, upon reflection, I did not feel au-
thorised to give it to her, and besides I thought
that if she died I should have to answer for it when
I returned, so, partly from conscience and partly
from prudence, I left her.

I may observe that this old woman was the only
sick person I ever saw in South America. The
temperate lives the people lead apparently give
them an uninterrupted enjoyment of health, and
the list of disorders with which the old world is
afflicted is altogether unknown. The beef on which
they almost entirely subsist is so lean and tough,
that few are tempted to eat more than is necessary,
and if a hungry Gaucho has swallowed too much
of a wild cow, the cure which nature has to perform is very simple. She has only by fever to deprive him of his appetite for a day or two, and he
is well again.

I have often remarked that the Gaucho has no
remedy for any small flesh wound, and does not
even keep the dirt from it, for his habit of body is
so healthy that the cure is positively going on as
he gallops along the plain.

*　*　*　*　*

THE PAMPAS.

*　　*　　*　　*　　*

I came to a post, and found horses in the corral, but the men all out. The woman told me they would be in soon, if I would wait. I saw a little child about seven years old, and said I would take him as a postilion. " Bién" (very well), said the woman, upon which the little boy was going to say something, but I took him by the arm, and leading him out to the corral, I caught our horses with a lasso which was lying on the ground.

After we had started, and had ridden about a league, " Oiga, Senor," said the little rosy-faced urchin, " yo no soy vaqueano" (I do not know the road.) I took up my whip and frightened him on before me ; but we were shortly overtaken by a man, who had galloped after us from the post as hard as his horse could go. He said he was the boy's father, that there were a number of " salteadores" (robbers) in the country—that it was not safe for the child, and that he had therefore come to conduct me. I had ridden more than a hundred miles, was very tired, not at all inclined to talk, and the man steadily galloped on before me. "Vea, Senor!" (see!) said the little boy as he frisked by my side, pointing to some wild ducks in a pool, which he wanted me to shoot at with my pistols.

His father was at this moment singing a wild sort of Spanish Hymn, and he had just got to the last note, upon which he was to hang for several seconds, when the merry little child, finding that there was no fun in me, loosened his rein, came up with his father, and gave his horse a blow as hard as he was able with the long whip which hung at his bridle, and then laughing, he darted away like a young colt, while his father with the greatest gravity continued the last note of his song.

*　　*　　*　　*　　*

I arrived for the night at a hut, where there were fowls, and I begged the woman to cook one of them immediately.

As soon as the water in a large pot had boiled, the woman caught a hen, and killed it by holding its head in her hand; and then, giving the bird two or three turns in the air, to my horror and utter astonishment, she instantly put the fowl into the pot, feathers and all ; and although I had resolved to rough it on my journey, yet I positively could not make up my mind to drink such broth or " potage au naturel" as I thought she was preparing for me. I ran to her, and, in very bad Spanish, loudly protested against her cookery; however, she quietly explained to me that she had only put the fowl there to scald it, and as soon as I let go her arm she took it out. The feathers all came off together, but they stuck to her fingers almost as fast as they had before to the fowl. After washing her hands, she took a knife, and very neatly cut off the wings, the two legs, the breast and the back, which she put one after another into a small pot with some beef suet and water, and the rest of the fowl she threw away,

*   *   *   *   *

THE PAMPAS INDIANS.

WHEN one compares the relative size of America with the rest of the world, it is singular to reflect on the history of those fellow-creatures who are the aborigines of the land; and after viewing the wealth and beauty of so interesting a country, it is painful to consider what the sufferings of the Indians have been, and still may be. Whatever may be their physical or moral character, whether more or less puny in body or in mind than the inhabitants of the old world, still they are the human beings placed there by the Almighty; the country belonged to them, and they are therefore entitled to the regard of every man who has religion enough to believe that God has made nothing in vain, or whose mind is just enough to respect the persons and the rights of his fellow-creatures.

A fair description of the Indians I believe does not exist. The Spaniards, on the discovery of the country, exterminated a large proportion of this unfortunate race; the rest they considered as beasts of burden, and during their short intervals of repose, the priests were ordered to explain to them, that their vast country belonged to the Pope at Rome. The Indians, unable to comprehend this claim, and sinking under the burdens which they were doomed to carry, died in great numbers. It was therefore convenient to vote that they were imbecile both in body and mind; the vote was seconded by the greedy voice of avarice, and carried by the artifices of the designing, and the careless indolence of those who had no interest in the question: it became a statement which historians have now recorded.

But although the inquiry has been thus lulled to rest, and is now the plausible excuse for our total ignorance on the subject, ought not the state of man in America to be infinitely more interesting than descriptions of its mines, its mountains, &c. &c. &c.

During my gallop in America, I had little time or opportunity to see many of the Indians; yet from what I did hear and see of them, I sincerely believe they are as fine a set of men as ever existed under the circumstances in which they are placed. In the mines I have seen them using tools which our miners declared they had not strength to work with, and carrying burdens which no man in England could support; and I appeal to those travellers who have been carried over the snow on their backs, whether they were able to have returned the compliment, and if not, what can be more grotesque than the figure of a civilized man riding upon the shoulders of a fellow-creature whose physical strength he has ventured to despise?

The Indians of whom I heard the most were those who inhabit the vast unknown plains of the Pampas, and who are all horsemen, or rather pass their lives on horseback.. The life they lead is singularly interesting. In spite of the climate, which is burning hot in summer, and freezing in winter, these brave men, who have never yet been subdued, are entirely naked, and have not even a covering for their head.

They live together in tribes, each of which is governed by a Cacique, but they have no fixed place of residence. Where the pasture is good there are they to be found, until it is consumed by

I

their horses, and they then instantly move to a
more verdant spot. They have neither bread, fruit,
nor vegetables, but they subsist entirely on the flesh
of their mares, which they never ride ; and the only
luxury in which they indulge, is that of washing
their hair in mare's blood.

The occupation of their lives is war, which they
consider is their noble and most natural employment ; and they declare that the proudest attitude of
the human figure is when, bending over his horse,
man is riding at his enemy.   The principal weapon
which they use is a spear eighteen feet long ; they
manage it with great dexterity, and are able to give
it a tremulous motion which has often shaken the
sword from the hand of their European adversaries.

From being constantly on horseback, the Indians
can scarely walk.   This may seem singular, but
from their infancy they are unaccustomed to it.
Living in a boundless plain, it may easily be conceived, that all their occupations and amusements
must necessarily be on horseback, and from riding
so many hours the legs become weak, which naturally gives a disinclination to an exertion which
every day becomes more fatiguing ; besides, the

pace at which they can skim over the plains on horseback is so swift, in comparison to the rate they could crawl on foot, that the latter must seem a cheerless exertion.

As a military nation they are much to be admired, and their system of warfare is more noble and perfect in its nature than that of any nation in the world. When they assemble, either to attack their enemies, or to invade the country of the Christians, with whom they are now at war, they collect large troops of horses and mares, and then uttering the wild shriek of war, they start at a gallop. As soon as the horses they ride are tired, they vault upon the bare backs of fresh ones, keeping their best until they positively see their enemies. The whole country affords pasture to their horses, and whenever they choose to stop, they have only to kill some mares. The ground is the bed on which from their infancy they have always slept, and they therefore meet their enemies with light hearts and full stomachs, the only advantages which they think men ought to desire.

How different this style of warfare is to the march of an army of our brave but limping, footsore men, crawling in the rain through muddy lanes, bending under their packs, while in their rear the mules, and forage, and packsaddles, and baggage, and waggons, and women—bullocks lying on the ground unable to proceed, &c. &c., form a scene of despair and confusion which must always attend the army that walks instead of rides, and that eats cows* instead of horses. How impossible would it be for an European army to contend with such an aerial force. As well might it attempt to drive the swallows from the country, as to harm these naked warriors.

A large body of these Indians twice crossed my path, as I was riding from Buenos Aires to Mendoza and back again. They had just had an engagement with the Rio Plata troops, who killed several of them, and these were lying naked and dead on the plain not far from the road. Several of the Gauchos, who were engaged, told me that the Indians had fought most gallantly, but that all their horses were tired, or they could never have been attacked : the Gauchos, who themselves

* On a long march it seldom happens that the bullocks are able to keep up with the men.

ride so beautifully, all declare that it is impossible
to ride with an Indian, for that the Indians' horses
are better than theirs, and also that they have such
a'way of urging on their horses by their cries, and
by a peculiar motion of their bodies, that even if
they were to change horses, the Indians would beat
them.   The Gauchos all seemed to dread very
much the Indians' spears.   They said that some of
the Indians charged without either bridle or saddle,.
and that in some instances they were hanging almost
under the bellies of their horses, and shrieking, so
that the horses were afraid to face them.   As the
Indians' horses got tired, they were met by fresh
troops, and a great number of them were killed.

To people accustomed to the cold passions of
England, it would be impossible to describe the
savage, inveterate, furious hatred which exists
between the Gauchos and the Indians.   The latter
invade the country for the ecstatic pleasure of murdering the Christians, and in the contests which
take place between them mercy is unknown.   Before I was quite aware of these feelings, I was
galloping with a very fine-looking Gaucho, who
had been fighting with the Indians, and after

listening to his report of the killed and wounded, I
happened, very simply, to ask him, how many pri-
soners they had taken? The man replied by a
look which I shall never forget—he clenched his
teeth, opened his lips, and then sawing his fingers
across his bare throat for a quarter of a minute,
bending towards me, with his spurs striking into
his horse's side, he said, in a sort of low, choking
voice, " Se matan todos," (we kill them all.) But
this fate is what the Indian firmly expects, and
from his earliest youth he is prepared to endure not
only death, but tortures, if the fortune of war
should throw him alive among his enemies; and
yet how many there are who accuse the Indians of
that imbecility of mind which in war bears the
name of cowardice. The usual cause for this
accusation is, that the Indians have almost always
been known to fly from fire-arms.

When first America was discovered, the Spaniards
were regarded by the Indians as divinities, and
perhaps there was nothing which tended to give
them this distinction, more than their possessing
weapons, which, resembling the lightning and the
thunder of Heaven, sent death among them in a

manner which they could not avoid or comprehend;
and although the Christians are no longer considered
as divine, yet the Indians are so little accustomed
to, or understood the nature of fire-arms, that it is
natural to suppose the danger of these weapons is
greater in their minds than the reality.

Accustomed to war among themselves with the
lance, it is a danger also that they have not learnt
to encounter; and it is well known that men can
learn to meet danger, and that they become familiar
with its face, when, if the mask be changed, and it
appear with unusual features, they again view it
with terror. But even supposing that the Indians
have no superstitious fear of fire-arms, but merely
consider their positive effects,—is it not natural
that they should fear them? In Europe, or in
England, what will people, with sticks in their
hands, do against men who have fire-arms? Why
exactly what the naked Indians have been accused
of doing—run away.—And who would not run
away?

But the life which the Indian leads must satisfy
any unprejudiced person that he must necessarily
possess high courage. His profession is War, his

food is simple, and his body is in that state of
health and vigour, that he can rise naked from the
plain on which he has slept, and proudly look upon
his image which the white frost has marked out
upon the grass without inconvenience. What can
we " men in buckram" say to this ?

The life of such a people must certainly be very
interesting, and I always regretted very much that
I had not time to throw off my clothes and pay a
visit to some of the tribes, which I should otherwise
certainly have done, as, with proper precautions,
there would have been little to fear ; for it would
have been curious to have seen the young sporting
about the plains in such a state of wild nature, and
to have listened to the sentiments and opinions of
the old ; and I would gladly have shivered through
the cold nights, and have lived upon mare's flesh
in the day, to have been a visitor among them.

From individuals who had lived many years
with them, I was informed that the religion of
the Pampas Indians is very complicated. They
believe in good spirits and bad ones, and they pray
to both. If any of their friends die before they
have reached the natural term of life, (which is very

unusual,) they consider that some enemy has prevailed upon the evil spirit to kill their friend, and
they assemble to determine who this enemy can be.
They then denounce vengeance against him. These
disputes have very fatal consequences, and have
the political effect of alienating the tribes from one
another, and of preventing that combination among
the Indians which might make them much more
dreaded by the Christians.

They believe in a future state, to which they
conceive they will be transferred as soon as they
die. They expect that they will then be constantly
drunk, and that they will always be hunting; and
as the Indians gallop over their plains at night,
they will point with their spears to constellations
in the Heavens, which they say are the figures of
their Ancestors, who, reeling in the Firmament, are
mounted upon horses swifter than the wind, and
are hunting ostriches.

They bury their dead, but at the grave they
kill several of their best horses, as they believe that
their friend would otherwise have nothing to ride.
Their marriages are very simple. The couple to
be married, as soon as the sun sets, are made to lie

on the ground with their heads towards the west. They are then covered with the skin of a horse, and as soon as the sun rises at their feet, they are pronounced to be married *.

The Indians are very fond of any sort of intoxicating liquor, and when they are at peace with Mendoza, and some of the other provinces, they often bring skins of ostriches, hides, &c., to exchange for knives, spurs, and liquor.

The day of their arrival they generally get drunk, but before they indulge in this amusement, they deliberately deliver up to their Cacique their knives, and any other weapons they possess, as they are fully aware that they will quarrel as soon as the wine gets into their heads.   They then drink till they can hardly see, and fight, and scratch, and bite, for the rest of the evening.   The following day they devote to selling their goods, for they never will part with them on the day on which they resolve to be tipsy, as they say that in that state they would be unable to dispose of them to advantage.

* I believe this would almost be a legal marriage in Scotland.

They will not sell their skins for money, which they declare is of no use, but exchange them for knives, spurs, maté, sugar, &c. They refuse to buy by weight, which they do not understand; so they mark out upon a skin how much is to be covered with sugar, or anything of the sort which they desire to receive in barter for their property. After their business is concluded, they generally devote another day to Bacchus, and when they have got nearly sober, they mount their horses, and with a loose rein, and with their new spurs, they stagger and gallop away to their wild plains.

Without describing any more of their customs, which I repeat only from hearsay, I must only again lament that the history of these people is not better known; for, from many facts which I heard concerning them, I really believe that they, as well as the Araucana Indians, possess many brave and estimable qualities. It is singular, however, to think how mutually they and the inhabitants of the old world are unacquainted with each other. These untamed soldiers know nothing of the governments, customs, habits, wants, luxuries, virtues, or follies, of our civilised world, and what does the civilised world know of them? It votes them savages *et voilà tout;* but as soon as fire-arms shall get into the hands of these brave naked men, they will tumble into the political scale as suddenly as if they had fallen from the moon ; and while the civilized world is watching the puny contests of Spaniards who were born in the old world, against their children who were born in the new one, and is arguing the cause of dependence *versus* independence, which in reality is but a quibble, the men that the ground belongs to will appear, and we shall then wonder how it is that we never felt for them, or cared for them, or hardly knew that they existed.

It may to many appear improbable that they should be ever able to overturn any of the feeble governments which at present exist; yet these men, without fire-arms, and with nothing in their hands but the lance, which is literally a reed, were twice within fifty leagues of Buenos Aires while I was in the country, and the Montaneros went among them while I was at San Luis, to offer to arm them. Besides this, the experience and history of the old world instruct us that the rise and fall of nations is a subject far beyond the scrutiny of man, and that, for reasons which we are unable to comprehend, the wild and despised tribes of our own world have often rushed from the polar towards the equatorial regions, and like the atmosphere from the north, have chilled and checked the luxury of the south; and therefore, however ill it may suit our politics to calculate upon such an event as the union of the Araucana and Pampas Indians, who can venture to say that the hour may not be decreed, when these men, mounted upon the descendants of the very horses which were brought over the Atlantic to oppress their forefathers, may rush from the cold region to which they have been driven, and with irresistible fury proclaim to the guilty conscience of our civilised world, that the hour of retribution has arrived; that the sins of the fathers are visited upon the children; that the descendants of Europeans are in their turn trampled under foot, and, in agony and torture, in vain are asking mercy from the *naked Indians?*

What a lesson this dreadful picture would afford!

However, it is neither my profession nor my wish to moralise; but it is impossible for a solitary individual to pass over the magnificent regions of America, without respecting the fellow-creatures who were placed there by the Almighty.

* * * * *

PASSAGE ACROSS THE GREAT
CORDILLERA.

THE mules were ordered at twelve o'clock, but
did not arrive till four: we had been waiting for
them with great impatience; at last we heard the
tinkling bell approaching, and they then came into
the yard of the Fonda (inn), driven by the capataz
and one peon.   The capataz was a tall stout man,
with a bad expression of countenance: we found
him cruel, lazy, insolent, cowardly, and careless of
everything but eating, and all this easily to be read
in his countenance.   The peon was a handsome,
slight-made, active young fellow.

There were sixteen mules of different sizes and
colours; they were all thin, but looked very
healthy and hardy.   One or two of them had
dreadful sore backs, which I pointed out to the
capataz, who promised to change them as soon as
he got out of Mendoza.   As my party consisted of
eight people, and as we had baggage sufficient for

six mules, we had only two spare ones, and these
unable to work; whereas I learnt afterwards, that
the capataz was bound to provide a much larger
proportion of extra mules, but he was as greedy
after lucre as he was after food, and to save a few
dollars he would have worked his poor mules to
death. However, I was then ignorant of the customs of the country, and indeed did not know
what was required for the journey I was about to
take; and anxious to be off, I ordered the mules to
be saddled.

As soon as this was done, the baggage-mules
were to be got ready. The capataz said he could
not load them, until every article of baggage was
brought into the yard, and accordingly he made a
great heap of it. He and the peon then divided it
into six parcels, quite different from each other in
weight or bulk, but adapted to the strength of the
different mules.

The operation of loading then began. The
peon first caught a great brown mule with his
lasso, and he then put a poncho over his eyes, and
tied it under his throat, leaving the animal's nose
and mouth uncovered. The mule instantly stood

still, while the capataz and peon first put on the large straw pack-saddle, which they girthed to him in such a manner that nothing could move it. They then placed the articles one by one on each side, and bound them all together, with a force and ingenuity against which it was hopeless for the mule to contend.

One could not help pitying the poor animal, on seeing him thus prepared for carrying a heavy load such a wearisome distance, and over such lofty mountains as the Andes; yet it is truly amusing to watch the nose and mouth of a mule, when his eyes are blinded, and his ears pressed down upon his neck in the poncho. Every movement which is made about him, either to arrange his saddle or his load, is resented by a curl of his nose and upper lip, which in ten thousand wrinkles is expressive, beyond description, of every thing that is vicious and spiteful: he appears to be planning all sorts of petty tricks of revenge, and as soon as the poncho is taken off, generally begins to put some of them into execution, either by running with his load against some brother mule, or by kicking him; however, as soon as he finds that his burden is not to be got

K

rid of, he dismisses, or perhaps conceals, his re-
sentment, and instantly assumes a look of patience
and resignation, which are really also the characteristics of his race, and which support them
under all their sufferings and privations.

As soon as the baggage-mules were ready, we
took up our pistols and carbines, and after mounting our mules, and shaking hands with the crowd
who had assembled in the yard, we bade adieu to the
Fonda of Mendoza. The last person that I bade
farewell to, was the old black cook, who was really
crying to see us go. She was one of the most
warm-hearted and faithful creatures I had ever met
with. She came to me just before I started, to
beg me to take care of myself, and she was then
half laughing and half crying. I was at the
moment going to throw away a pair of green
goggle-spectacles, with shining, lackered rims,
which I had bought to cross the snow of the Cordillera, but which I had just condemned as troublesome and useless; however, seeing the old woman's
grief, I gave them to her, and put them on the
bridge of her short black nose, sticking the ends of
them into her woolly hair. She considered it,

perhaps, as an act of kindness, and began to cry; and although the group around us were roaring with laughter, the spectacles remained on her nose all the time I was conversing with her. She then took them off, and looking at them with great pride and delight, put them into the bosom of her gown.

The saddling of the mules had taken up so much time that the sun had nearly set. It was still oppressively hot; however, the siesta, which with eating, &c., is in Mendoza an operation of six hours, was over, and the people were standing at their doors to see us pass; but as we went by the Almeida road, we soon got out of the town. In the stream which runs along the row of poplars which shade this Almeida, or public walk, the people were bathing as usual, without any dresses, and apparently regardless of each other. The young called out to us, and many jokes were taken and given.

After passing the long Almeida, the road, for about two leagues, passes through a country artificially irrigated from the Rio de Mendoza, and its luxuriance and fertility are quite extraordinary. The brown mud walls which bound the road were

covered with grapes, which hung down in beautiful
clusters; and the number of peach-trees, loaded
with fruit, and scattered among rich crops of corn
and other agricultural produce, gave the scene an
appearance of great luxury and abundance; while
the mountains of the Cordillera formed a magnificent boundary to a picture which, to one about
to cross the Andes, is peculiarly interesting.   As
soon as the boundary of irrigation is passed, the
country ceases to be productive.   The soil, light
and sandy, produces no sort of herbage, and for
more than thirty miles, the road, as it approaches
the mountains, passes through a plain, which bears
nothing but low stunted shrubs; and when one considers that such has probably been its produce since
the creation of the world, it is surprising to see that
vegetation, so nearly extinct, should have lingered
so long without expiring.   However, its existence
in these plains proves that they are capable of producing crops for man, whenever his industry will
seek for the treasure.

The road across this flat country is always te-
dious; for the mountains, on leaving Mendoza,
appear within three or four miles of the town, and

the path seems literally to lengthen as one goes. We found it particularly so, as we had to travel in a night which was unusually dark. The plain before us was not visible, while the black outline of the mountains against the sky appeared close to us, or rather immediately above us. However, we at last got to the first ravine of the Cordillera; and then, with the noble mountains towering over our heads, sometimes lost in darkness, and sometimes marked out by the few stars which were visible, we followed the sound of the water, until the distant light at the post-hut, and the barking of the dogs as they came rushing towards us, told us that we should now cross the stream, which we did, and then rode up to the post. The dogs continued barking, and occasionally biting at our mules' tails, until the postmaster and another man came to us. They were sleeping by the embers of a fire, in the kitchen or shed which was before us. One side was completely open, the other three were of boughs wattled, but so open that the smoke easily escaped.

The post of Villa Vicencia, which in all the maps of South America looks so respectable, now

consists of a solitary hut without a window, with a
bullock's hide for a door, and with very little roof.
As the night was cold, I preferred sleeping in the
kitchen by the fire, leaving the mules to do as
they chose, and to go wherever their fancies might
incline them. I took for my pillow one of the
horses' skulls, which in South America are used as
chairs, and wrapping myself up in my poncho, I
dropped off to sleep. When I awoke, which was
before daybreak, I found two peons and one of my
party asleep round the fire, and a great dog snoring
at my back.

I called out for the capataz, who came to me
rubbing his eyes, and looking dirty and sleepy,
and I told him to go after his mules; but one of
the men said that the peon was already gone. Our
men were also up, preparing some soup, and as the
day began to dawn, and as the mules did not
appear, I resolved to find out the baths, which are
about a mile off. I followed a path until I came
to a spot where I was surrounded by hills, which it
seemed quite impossible to climb even on hands
and knees; however, on proceeding, I found a singular passage cut in the rock, and climbing up to

it, came suddenly to a little spot, in which were
the ruins of two or three huts, and three or four
tents.

The huts and the tents were swarming with
people, and the discovery of twenty or thirty fellow-creatures in such a sequestered spot was altogether unexpected. They had come there from
great distances for the purpose of bathing, and
many of them I afterwards learnt were very respectable people. As I had no time to lose, and
wanted to bathe, I asked a man who was looking
out of a tent, where the baths were? With the
indifference and indolence usual in the country, he
made no reply, but he pointed with his chin to
some little walls close before him, two or three feet
high, built with loose stones and in ruins. I was
also close to them, so I took off my jacket and my
belt of pistols, and walked towards them; but not
believing they could be baths, I looked towards the
man, and asked him if they were there. He made
with his head the usual sign of " Si;" so I walked
towards the walls, and to my astonishment I found
a hole a little bigger than a coffin, with a woman
lying in it! Seeing that there was no room for me

there, I reconnoitred the spot, and found another hole about ten yards above the lady, and another about the same distance below her. As the water ran from the one to the other, I thought I might as well act the part of the wolf as be the lamb, and I therefore went up the stream, and got into the upper bath. I found the water very hot and agreeable; and without troubling myself about its analysis, drank some from the spot where it issued from the ground, and feeling that I had then given it a fair trial, I set off to return. In passing the huts and the tents I looked into them;—they were crowded with men, women, and children, of all ages, and mingled together in a way which would not altogether be admitted at our English bathing-places; but among the Andes customs and ideas are different, and if a lady has there the rheumatism, she sees no harm in trying to wash it away by the waters of Villa Vicencia.

As soon as I got back to the post-hut I found the mules all saddled; so, after drinking some soup and eating a piece of the hind-leg of a guanaco, I set off for Uspallata, where it was proposed we should sleep.

The road, on leaving Villa Vicencia, instantly turns up a ravine, which is one of the finest passes in the Cordillera. The mountains are extremely steep on both sides, and, as the ravine winds in many directions, one often comes to a spot which has the appearance of a Cul-de-Sac, from which there is no exit to be seen. In some places the rock hangs perpendicularly over-head, and the enormous fragments which nearly block up the road, contrasted with those which appear to be on the point of falling, add to the danger and to the grandeur of the scene. As we were passing we saw a guanaco on the very highest summit of one of the mountains. He was there evidently for safety; and as he stood against the blue sky, his attitude, as he earnestly watched us, was very expressive of his wild free life; and his small head and thin neck denoted the speed with which he was about to save himself.

I had ridden on by myself about fifteen miles, and had gained, by a constant ascent, the summit of the Paramillo, the high range of mountains which overhang Villa Vicencia. The view from this point is very interesting. The ground continues level

for a short distance, and then rapidly descends
towards the valley of Uspallata, which is about
thirty miles off.

This valley is the upper base of the great range
of the Cordilleras; and it is, at first, surprising to
see that the hills of the Paramillo, which had
appeared so lofty, are very humble features compared with the stupendous barrier which, in spite
of its distance, appears to be on the point of
obstructing the passage.

This enormous mass of stone, for it appears to
be perfectly barren, is so wild and rude in its
features and construction, that no one would judge
that any animal could force its way across the summit, which, covered with snow, in some places
eternal, seems to be a region between the heavens
and the practicable habitation of man; and indeed
to attempt to pass it, except by following up in a
ravine the course of a torrent, would be altogether
impossible.

From the Paramillo, the view on the east, or
contrary direction, is also very interesting. It is
pleasing to look down on the difficulties which
have been surmounted even to gain this point; and

beyond is a vast expanse of what, at first, very much resembles the ocean, but which one soon recognises as the vast plains of Mendoza and the Pampas.

The natural exhalation from the earth covers them with a cloud of uncertainty : places which one has heard talked of as points of importance are lost in space, and the hopes and passions and existence of mankind are buried in the atmosphere which supports them. But one has not much time for moralizing on the summit of the Paramillo, for it is such a windy spot, that a man's most rational exertion there is to hold on his hat; and as the large broad-brimmed one, which I had purchased at Mendoza, made several attempts to return there, I and my mule proceeded towards the valley of Uspallata. After going a league or two, I observed on both sides of me large tawny-coloured fungus-looking substances, which in size, shape, and colour, so resembled lions lying on the ground, that sometimes I really could not distinguish whether they were or not.

In the Pampas I had always observed the singular manner in which all animals, particularly birds,

are there protected from their enemies by plants or
foliage which resemble them; and as I knew there
were a great number of lions about Villa Vicencia,
and could see the track of their large feet in my
path, I began to think that some of them were
really lying before me. However, it seemed foolish
to stop, and therefore I continued for some time; at
last, coming to a small coppery vein in the rock, I
thought it would be a good excuse to inspect it, so
I remained there cracking the stones till two of my
party came up, and their first observation to me
was, how very like the substances around us were
to lions.

One of the party had a horse's leg in his hand.
He told me that he had never been so tired in his
life; that his mule, in mounting the hill, had become quite exhausted; and that, when he got off to
lead her, she would not follow him: that, in despair, he made her drink up his flask of brandy,
and that then, taking as a whip a dried-up horse's
leg that was lying on the ground, he remounted
the mule, which had gone very well ever since;
" But, Sir," said my honest companion, " whether
it be the brandy that has got into her head, or the

notion of being beaten with a horse's leg that has urged her on, I cannot tell you."

We continued our course together, and descending the hill, came to the district in which the Uspallata Mines are situated. The climate of the country in which these mines are situated is what would naturally be expected from its latitude and elevation. The former places it under a hot sun, the latter imparts to it a considerable degree of cold; and as the air is both dry and rarefied, there is little refraction, and consequently the heat and light of day vanish almost as soon as the sun is below the horizon. In visiting these mines in winter we found the days hotter than the summer in England, when at night the water constantly froze hard by our sides as we slept crowded together in the small hut. The whole of the country is the most barren I ever witnessed, and from this singular cause, that it never rains there*.

---

* Without attempting to explain the cause of this phenomenon, the following are some of the facts on which the statement is founded:—

1. The huts at several of the mines are built exactly across the ravine, in such a manner, that if water was ever to come

The soil consists of the decomposed rock, which remains on the steep surface of the mountain, and rolls from under the foot like the loose cinders of Etna and Vesuvius: there is no herbage of any sort or kind upon it. A few low, resinous shrubs are scattered about ; but, from the severity of the climate, in most places they grow along the ground. The dead animals which are lying about are all dried up in their skins, and have a most singular appearance: indeed the whole scene is a very striking example of what a desert the earth would be without water. One of the Cornish miners, after gazing about him with astonishment, took up a handful of the green barren soil, and looking into it with great attention, he said, " Why, surely there must be poison in this ground !"

down the ravine, it must necessarily pass through the huts, or over them.

2. One of the lodes runs up the bottom of a ravine, and the old shafts which are formed in it are in the natural drain of the ravine. These shafts at bottom are dry, and have no appearance of having contained water.

3. The miner, who, to keep possession of the mines, had lived there alone for two years, told us, that during that time it had not rained once.

We had scarcely passed the mines when the sun set, and although we saw the post-hut of Uspallata, yet we had great difficulty in reaching it. The rest of the party were lost, and did not arrive till midnight. My first object was to get something for our poor mules; there was very little in the plain except hot stones and resinous shrubs, but I learnt from the man that he had a potrero (or enclosed field) full of grass: he began a long story about how much I was to pay—however, I cut him very short, and sent him off with the mules, who, poor things, were no doubt delighted with their unexpected supper.

We then earnestly inquired of the man what he had for us to eat? And as we all three stood round him, our earnest and greedy looks were an amusing contrast to the calm tranquillity with which he replied " No hai," to everything we asked for; at last we found out that he had got dry peaches and live goats. We put some of the former into a pot to boil, and in process of time the boy, who was sent out on horseback with a lasso to catch a goat, arrived. The little fellow could not kill it, and the man was gone for wood;

so partly to put an end to the animal's fears, and partly because I was very hungry, I put a pistol to his ear, and in a short time he was roasting on the burning embers.

At this moment an English lady, a child about seven years old, two or three younger ones, and a party of peons arrived. They had, with no other protection, passed the Cordillera, and had ridden for twelve or fourteen hours that day in order to get to Uspallata.

The situation of a country-woman with a family of little children interested us very much, and it was pleasing to hear that they had crossed the Cordillera without any accident. The eldest child, who was a very fine boy, had ridden the whole way, but the other little chubby-faced creatures had each been carried upon a pillow in front of the peons' saddles.

In the history of the hut of Villa Vicencia I had often heard that, in spite of its desert situation and want of comfort, an English lady, who was passing with her husband to Chili about seven or eight years ago, had been confined there, and had remained there until she and her little infant were capable of prosecuting their perilous journey; and when I saw the wretched abode, I had often felt how cheerless it must have been for her to have remained there so long.

The lady who now came to Uspallata was the very person whose singular sufferings I have described, and the fine little boy was the child that was born at Villa Vicencia. He had been in Chili ever since, and now the little manly fellow had ridden across the Cordillera, and was about to introduce his brothers and sisters to the wild hovel in which it had been decreed that he should be born.

In the morning, before daybreak, we made preparations for starting. Some part of the goat was to form our breakfast; we had some tea with us, and I was very anxious to get some milk, but when I asked the man, he replied, " Leche no hai," with a look that seemed to doubt there being any in the universe. The cows, he said, were four leagues off, and that they would not come for a couple of hours. " Have the goats no milk?" asked I; the fellow laughed at the idea; however I found out that they had kids, and I therefore insisted on his

L

sending the boy for a she-goat. This order was complied with, and in a short time the boy came, dragging a poor creature with his lasso. She was altogether scared, and was leaping and jumping to get away; however, our peons helped, and she was thrown down upon her side. One peon knelt upon her head, and one of our men held her hind legs, while the boy milked her on one side, and then turning her round, in spite of her struggles, she was milked on the other side. They then let her go, and happy was she at regaining her liberty, after being scared at the uncouth operation she had just undergone.

The mules were now nearly laden, when one of the Cornish miners told me that the capataz wanted to put baggage upon the mule which had got a sore back, and which, according to his agreement, he ought to have changed at Mendoza. I instantly went to the capataz, and found him with his long knife in his hand, actually cutting the poor creature's back, preparatory to putting on the pack-saddle. I told him to desist, but he was explaining to me how he was going to place the saddle, so that it should not hurt the mule, and he was just going to put on a small straw-pad, when I at once put an end to the argument. As soon as the baggage was ready, we threw upon it two or three dead sheep, and, in quitting Uspallata, took leave of the last inhabited hut on the east side of the Cordillera.

I was steadily riding my mule at the rate of five miles an hour, in order to measure by my watch the breadth of the plains of Uspallata, when we met an old Gaucho huntsman, with two lads, and a number of dogs, which at once put a stop to my calculation. He had several loose horses, over one of which was hanging the carcass of a guanaco.

He had been hunting for lions, and had been among the mountains for two days, but had had little sport. The Gaucho was a fine picture of an old sportsman. Round his body were the " bolas" (balls), which were covered with clotted blood. His knees were admirably protected from the bushes by a hide which was under his saddle, and which in front had the appearance of gambadoes.

He was mounted on a good horse, his lasso in coils hung at his saddle. As soon as we stopped, he

was surrounded by his dogs, which were a very odd
pack.   Some of them were very large, and some
quite small, and they seemed to be all of different
breeds ; many had been lamed by the lions and
tigers, and several bore honourable scars.   I regretted very much indeed that I had not time to
follow the sport, which must have been highly
interesting.

As soon as the dogs unkennel a lion or a tiger,
they pursue him until he stops to defend himself.
If the dogs fly upon him, the Gaucho jumps off his
horse, and while the animal is contending with his
enemies, he strikes him on the head with the balls,
to which an extraordinary momentum can be given.
If the dogs are at bay, and afraid to attack their
foe, the Gaucho then hurls the lasso over him, and
galloping away, he drags him along the ground,
while the hounds rush upon him and tear him.

The mountains now seemed to be really over our
heads, and we expected that we should have immediately to climb them, but for many hours we went
over a plain as dry and barren as the country
already described on the other side of Uspallata,
and which wound its course among the mountains.

At last we crossed a rapid torrent of water, and then immediately afterwards came to another, which takes its rise at the summit of the Andes, and whose course and comparatively gradual descent directs the passage ; and it is on this spot that the traveller may proudly feel that he is at last buried among the mountains of the Andes. The surface of the rocks which surrounded us afforded no pasture, and the gnarled wood and the stunted growth of the trees announced the severity of the climate in winter ; yet the forms of the mountains, and the wild groups in which they stood towering one above another can only be viewed with astonishment and admiration.

Although the sun was retiring, and the mules very tired, we wished to have gone on half an hour longer, but the peon assured us we should not find so good a place, and, pointing to some withered herbage, and some large loose stones, he earnestly advised me to stop, saying, " Hai aqui pasto bueno para las mulas, y para su merced buen alojamiento, hai agua, aqui hai todo" (here is pasture for the mules, and for your excellency good lodging, water, and everything.) We therefore dismounted

near a spring, and having collected wood, and the
miners having cooked our supper, we lay down on
the ground to sleep. The air was cool and refreshing, and the scene really magnificent.

As I lay on the ground upon my back, the objects around me gradually became obscure, while
the sun, which had long ago set to us, still gilded
the summits of the highest mountains, and gave
a sparkling brightness to the snow which faded
with the light of day. The scene underwent a
thousand beautiful changes; but when it was all
lost in utter darkness, save the bold outline which
rested against the sky, it appeared more beautiful
than ever.

The peon, who was always very active, was up
long before day-break, and we were awakened by
the bell-mule and the others which were now collected. We got up in the dark, and as our party
were preparing to start, the group, though indistinctly seen by the blaze of the fire, was a very
odd one. The three miners were eating their
breakfasts seated on loose stones round a large fragment of rock, which served as a table. Their
elbows were squared, and they were eagerly bend-

ing over the food before them. The peons, with their dark brown faces, and different coloured caps, handkerchiefs, and ponchos, were loading the " carga" mules. Some of the party were putting on their spurs ; others were arranging their toilette. The light was now faintly dawning on the tops of the highest mountains, and the snow was just dis- covered lying in large patches and ridges. The bottoms of the ravines were in dark shade, and white windy clouds were flying across the deep blue sky—for some moments all was silent : however, as soon as the mules were ready we mounted, and we were off before we could distinctly see ; but the mules picked their way, and continually ascending by a path covered with great stones, and imprac- ticable to any animal except a mule, we continued to follow the course of the great stream, which was a torrent, roaring and raging, and altogether im- passable.

The sufferings of the poor mules now attracted our attention ; they had travelled from Mendoza with but little rest, and little food ; still they required no driving, but were evidently making every possible exertion to keep up with the mule which carried the bell. Occasionally the " carga" would require adjusting, and the peon, throwing his poncho over the creature's eyes, would alter it, while the rest continued their course, but the poncho was no sooner removed than the mule, trotting and braying, joined the troop, never stopping till he came to the bell.

On the road, the number of dead mules, which indeed strew the path from Mendoza to Santiago, seemed to increase, and it was painful to see the living ones winding their path among the bones and carcasses of those who had died of fatigue. By the peculiar effect of the climate, most of these poor creatures were completely dry, and as they lay on the road with their hind legs extended, and their heads stretched towards their goal, it was evident from their attitudes that they had all died of the same complaint—the hill had killed them all.

After passing one or two very rapid torrents, we came to a mountain which was one precipitous slope from the top to the torrent beneath. About half way up, we saw a troop of forty guanacos, who were all gazing at us with great attention. They were on a path, or track, parallel to the water, and as the side of the mountain was covered with loose stones, we were afraid they would roll some of them down upon us.

On the opposite side of the water, was one of the most singular geological formations which we had witnessed. At the head of a ravine was an enormous perpendicular mountain of porphyry, broken into battlements and turrets, which gave it exactly the appearance of an old castle, on a scale, however, altogether the subject of a romance. The broken front represented, in a most curious manner, old fashioned windows and gates, and one of the Cornish miners declared " he could see an old woman coming across a draw-bridge."

As I was looking up at the region of snow, and as my mule was scrambling along the steep side of the rock, the capataz overtook me, and asked me if I chose to come on, as he was going to look at the " Ladera de las Vaccas," to see if it was passable, before the mules came to it *. He accordingly trotted on, and in half an hour we arrived at the

* When first, from the melting of the snow, the Cordillera is " open," this passage is always impassable; but it becomes broader towards the end of summer.

spot. It is the worst pass in the Cordillera. The mountain above appears almost perpendicular, and in one continued slope down to the rapid torrent which is raging underneath. The surface is covered with loose earth and stones which have been brought down by the water. The path goes across this slope, and is very bad for about seventy yards, being only a few inches broad; but the point of danger is a spot where the water which comes down from the top of the mountain either washes the path away, or covers it over with loose stones. We rode over it, and it certainly was very narrow and bad. In some places the rock almost touches one's shoulder, while the precipice is immediately under the opposite foot, and high above the head are a number of large loose stones, which appear as if the slightest touch would send them rolling into the torrent beneath, which is foaming and rushing with great violence. However, the danger to the rider is only imaginary, for the mules are so careful, and seem so well aware of their situation, that there is no chance of their making a false step. As soon as we had crossed the pass, which is only seventy yards long, the capataz told me that it was a very bad

place for baggage-mules, that four hundred had
been lost there, and that we should also very probably lose one; he said, that he would get down to
the water at a place about a hundred yards off, and
wait there with his lasso to catch any mule that
might fall into the torrent, and he requested me to
lead on his mule. However, I was resolved to see
the tumble, if there was to be one, so the capataz
took away my mule and his own, and while I stood
on a projecting rock at the end of the pass, he
scrambled down on foot, till he at last got to the
level of the water.

The drove of mules now came in sight, one following another; a few were carrying no burdens,
but the rest were either mounted or heavily laden,
and as they wound along the crooked path, the
difference of colour in the animals, the different
colours and shapes of the baggage they were carrying, with the picturesque dress of the peons, who
were vociferating the wild song by which they drive
on the mules, and the sight of the dangerous path
they had to cross,—formed altogether a very interesting scene.

As soon as the leading mule came to the com-

mencement of the pass, he stopped, evidently unwilling to proceed, and of course all the rest stopped also.

He was the finest mule we had, and on that account had twice as much to carry as any of the
others; his load had never been relieved, and it consisted of four portmanteaus, two of which belonged
to me, and which contained not only a very heavy
bag of dollars, but also papers which were of such
consequence that I could hardly have continued
my journey without them. The peons now re-
doubled their cries, and leaning over the sides of
their mules, and picking up stones, they threw them
at the leading mule, who now commenced his journey over the path. With his nose to the ground,
literally smelling his way, he walked gently on,
often changing the position of his feet, if he found
the ground would not bear, until he came to the
bad part of the pass, where he again stopped, and
I then certainly began to look with great anxiety
at my portmanteaus; but the peons again threw
stones at him, and he continued his path, and
reached me in safety; several others followed. At
last a young mule, carrying a portmanteau, with

two large sacks of provisions, and many other
things, in passing the bad point, struck his load
against the rock, which knocked his two hind legs
over the precipice, and the loose stones immediately
began to roll away from under them: however his
fore-legs were still upon the narrow path; he had
no room to put his head there, but he placed his
nose on the path on his left, and appeared to
hold on by his mouth: his perilous fate was soon
decided by a loose mule who came, and in walking
along after him, knocked his comrade's nose off the
path, destroyed his balance, and head over heels
the poor creature instantly commenced a fall which
was really quite terrific. With all his baggage
firmly lashed to him, he rolled down the steep
slope, until he came to the part which was perpendicular, and then he seemed to bound off, and
turning round in the air, fell into the deep torrent
on his back, and upon his baggage, and instantly
disappeared. I thought, of course, that he was
killed; but up he rose, looking wild and scared,
and immediately endeavoured to stem the torrent
which was foaming about him. It was a noble effort; and for a moment he seemed to succeed, but

the eddy suddenly caught the great load which was upon his back, and turned him completely over; down went his head with all the baggage, and as he was carried down the stream, all I saw were his hind quarters, and his long, thin, wet tail, lashing the water. As suddenly, however, up his head came again; but he was now weak, and went down the stream, turned round and round by the eddy, until, passing the corner of the rock, I lost sight of him. I saw, however, the peons, with their lassos in their hands, run down the side of the torrent for some little distance; but they soon stopped, and after looking towards the poor mule for some seconds, their earnest attitude gradually relaxed, and when they walked towards me, I concluded that all was over. I walked up to the peons, and was just going to speak to them, when I saw at a distance a solitary mule walking towards us!

We instantly perceived that he was the Phaeton whose fall we had just witnessed, and in a few moments he came up to us to join his comrades. He was of course dripping wet; his eye looked dull, and his whole countenance was dejected: however, none of his bones were broken, he was very

little cut, and the bulletin of his health was altogether incredible.

With that surprising anxiety which the mules all
have to join the troop, or rather the leading mule
which carries the bell, he continued his course, and
actually walked over the pass without compulsion,
although certainly with great caution.

We then continued our course for two hours, until we came to the " Rio de las Vaccas," which is
the most dangerous torrent of any of those which
are to be crossed. We got through it with safety,
but it was very deep, and so excessively rapid, that
large stones were rolled down it with the force
of the water. The mules are accustomed to these
torrents, but they are, notwithstanding, much
frightened at them, and it is only long spurs that
can force them into them.

While we were crossing, the peons stood down
the stream, with their lassos hurling round their
heads, in order to catch anything which might have
been carried away ; but as the boxes which I had
seen washed from the mules were dashed to pieces
before they had got twenty yards, the peon's lasso
came a little too late ; and besides this, as the

mule is their own property, I used sometimes to think that, in the hurry and indecision of the moment, they would probably catch him instead of the rider.

When a large party cross this river, and when it is deep, it is really amusing, after one has got across it, to observe the sudden change of countenance of one's friends as *they* ride through it; sometimes perched up on the top of a fragment of rock barely covered, and expecting the next step to be their last; and sometimes scrambling out of a hole, with uplifted eye-brows, open mouth, and an earnest expression of uneasiness and apprehension—and these are really situations into which the traveller in the Andes is often thrown, though they disconcert the gravity and solemnity of his " Personal Narrative."

After passing the Rio de las Vaccas, the ravines appear to grow narrower and steeper, and the tops of the mountains, which are those of the highest range, are rugged, with sharp edges and pinnacles.

We here came to a quantity of snow and rubbish, which had been washed down, and which we had great difficulty to pass, for it occasionally broke under the weight of the mules, who recovered themselves in a surprising manner, and as if accustomed to it.

We now passed one of the brick huts, which, at every two or three leagues, have been built to protect the traveller from the dreadful storms which here assail him, and after continuing our course till the sun was low, we stopped at the second of these huts.

We saw a party of loose mules at some distance standing among the stones; and leaving my mule at the hut, I walked to them, and found two or three " arrieros" on the ground asleep.

I leaned over one fat fellow, and asked him to give me something to eat, for we had lost all our provisions at the Ladera de las Vaccas. As he awoke, he seemed at first alarmed at seeing a stranger well armed so near him; however, we soon came to an understanding, and in a few seconds he was putting some money into a long purse, while I was walking towards the hut, with my arms filled with hard sea biscuits, some dried beef (charque) with one hand full of salt, and in the other red Chili pepper.

With this our men prepared a good dinner, while I reconnoitred our situation. It was barren

M

and desolate beyond description; and the mules, now unsaddled, were standing in the attitudes in which they had been unladen—their heads were nodding, or drooping, and they were putting up their backs and going to sleep, which was the only comfort they could enjoy, for there was literally nothing for them to eat.

The snow was all around us, and the features of the scene so large, that one could not but reflect on the situation of the many travellers, who, in these parts of the Andes, have been overtaken by the storm, and have perished.

The capataz told me that these " temporales" are so violent that no animal can live in them; that there is no warning, but that all of a sudden the snow is seen coming over the tops of the mountains in a hurricane of wind; that hundreds of people have been lost in these storms; that several had been starved in the house before us; and that only two years ago, the winter, by suddenly setting in, as it generally does, had shut up the Cordillera, and had driven ten poor travellers to this hut. When the violence of the first storms had subsided, the courier came to the spot, and found six of the ten lying dead in the hut, and by their sides the other four almost dead with hunger and cold. They had eaten their mules and their dog, and the bones of these animals were now before us.

These houses are all erected upon one plan, and are extremely well adapted to their purpose. They are of brick and mortar, and are built solid, ten or twelve feet high, with a brick staircase outside. The room which is on the top of this foundation, in order to raise it above the snow, is about twelve feet square; the walls are extremely thick, with two or three small loop-holes about six inches square; the roof is arched, and the floor is of brick.

A place so small, of so massive a construction, necessarily possesses the character of a dungeon; and, as one stands at the door, the scene around adds a melancholy gloom to its appearance, and one cannot help thinking how sad it must have been, to have seen the snow, day after day, getting deeper and deeper, and the hope of escaping hourly diminishing, until it was evident that the path was impracticable and that the passage was closed! But without these reflections, the interior is melancholy enough.

The table, which had been fixed into the mortar, was torn away; and to obtain a momentary warmth, the wretched people who had been confined here had, in despair, burnt the very door which was to protect them from the elements. They had then, at the risk of their lives, taken out the great wooden lintel, which was over the door, and had left the wall above it hanging merely from the adhesion of the mortar. This operation had evidently been done with no instrument but their knives, and it must have been a work of many days.

The state of the walls was also a melancholy testimony of the despair and horror they had witnessed. In all the places which I have ever seen, which have been visited by travellers, I have always been able to read the names and histories of some of those who have gone before me; for when a man has nothing to lament, but that his horses have not arrived, or in fact that he has nothing to do, the wall appears to be a friend to whom many intrust their names, their birth-places, the place they propose to visit, and sometimes even the fond secrets of their hearts; but I particularly observed that, in these huts on the Andes, not a name was to

be seen, or a word upon the walls. Those who
had died in them were too intent upon their own
sufferings; the horror of their situation was un-
speakable, and thus these walls remain the silent
monuments of past misery.

As the air was very cold, and the wind very
high, we slept in this hut, and before day-break
we were once again upon our poor jaded mules, in
order to cross the Cumbre, while the surface of the
snow was hard from the night's frost. After climbing a little but very steep hill, we came upon a small
flat landing-place, which was the most dreary looking spot I think I ever saw. I asked the peon
what the wooden cross before us meant? After
looking over each of his shoulders. he told me that
the spot for many years was haunted by the ghost
of a mulish-looking sort of man who used to terrify
all the arrieros and peons who passed, and that
they, therefore, had been absolutely obliged to get
a priest to put up the cross before us. "And has
that driven the ghost away?" said I, laughing.
"Si," said the peon, with a look of confidence and
courage which had rather deserted his face while he
was describing the shape of the spectre; and he

then assured me with great earnestness, " that now he was never seen, and that I need not be afraid."

The torrent which we had so long followed, now turned up the ravine to the right. We had pursued it from the east towards the west, but our path was now obstructed by the Cumbre, or upper ridge of the Cordillera, which no artifice can avoid, and which is a mountain covered with loose, decomposed rock, at an angle of very nearly forty-five degrees. At the foot is another of the huts, without door, table, or lintel, and in which many people have died.

After resting my mule for a short time, and then girthing my saddle as tight as possible, during which operation he was always trying to bite me, I whispered a little comfort into his long ear; I mounted, and then squaring my shoulders and giving a kick or two with my spurs, I commenced the climb, followed by the party of riders and carga mules.

The path ascended in zig-zags from the bottom to the top, and the whole time I was obliged to hold on by the thin mane of the mule. The turnings were so short, that the animal was almost falling backwards; however, on he went, with a determination and patience that was quite astonishing. At times he stopped, but the path was so steep, and the decomposed rock so loose, that of his own accord in a few seconds he continued. It was very picturesque and interesting to see the whole party beneath, threading their way in different paths above each other; some going towards the north, and others towards the south—to see the riders leaning forwards, every animal straining to his utmost, and to hear the peons cheering on their mules by a song which was both wild and melodious.

After climbing in this singular manner for about an hour, I reached the summit, and it was really a moment of great triumph and satisfaction. Hitherto I had always been looking upwards, but now the difficulties were all overcome, and I was able to look down upon the mountains. Their tops were covered with snow; and as the eye wandered over the different pinnacles, and up the white trackless ravines, one could not but confess that the scene, cheerless and inhospitable as it was, was nevertheless a picture both magnificent and sublime.

Proceeding among some broken ground along
the summit, I saw a very large wooden cross, which
I rode up to. It was supported by a heap of stones
piled round the bottom of it, but it did not stand
perpendicular. It was roughly hewn, morticed
together, and fixed by a large spike nail, which had
rusted the wood, and being loosely clinched, the
cross creaked with the wind. There was a rough
inscription, cut out with a knife, along the bar of
the cross; but it was so much above my head, and
so bleached by the weather, that I could not read
it. In the wild desolate situation in which it stood,
it certainly looked very appropriate and interesting,
and I stood at the foot of it leaning over my mule
until the party came up; and then the peon told
me that it was placed there by two arrieros to
commemorate the murder of their friend. Thus
reminded that we had not yet risen above the bad
passions of man, it was painful to see the emblem
of his hopes standing as the monument of his
guilt!

We now found it extremely cold; the snow was
very deep, and the mules' path a most extraordinary
one   A deep narrow passage had been cut by the

constant travelling of these animals, but the wall of
snow on each side obliged the rider to put his feet
on the mule's ears; besides this, as they always
tread on the same spot, every step was into a hole
which was often above their knees. On the snow
there was a great deal of blood from mules which
had gone before, and it was only astonishing that
they could proceed at all.

"What a magnificent view!" said I to one of
my companions, whose honest heart and thoughts
were always faithful to old England. "What
thing can be more beautiful?" I added. After
smiling for some seconds, he replied, "Them things,
sir, that do wear caps and aprons."

After descending about a mile with great trouble
and difficulty, we came to another of the huts, which
was in the same state as all the rest, but surrounded
by about twelve feet of snow; for on the Chili side
of the Andes there is always much more snow than
on the other. After passing this house we resolved
to quit the path, which was getting more bloody
and more difficult, and we attempted to take a
nearer cut by riding over the snow, which was
everywhere very deep. It bore us very well for

some time; but as we got lower down, and as the
heat of the day increased, our mules began to sink
into it: however, they managed to regain the path,
except the poor brown mule who was carrying the
four heavy portmanteaus. He had hitherto surmounted every difficulty, and with a healthy eye
and a patient countenance had always led the way;
but now his treacherous path was breaking under
him, and after floundering on in a most extraordinary manner, literally raising himself by his nose,
he could proceed no farther, and the portmanteaus
at his side all rested on the snow. Before this the
capataz and peon had only cheered him by their
voices, but they now went to his assistance. They
lifted up his two fore-legs out of the holes which
they had made, and they put them on the surface
of the snow. They then went on each side, and
with one hand on his tail and the other under his
belly, the poor creature rose. The two men then
instantly jumped behind the mule, and with their
hands over their heads they both held the mule's
tail, pushing it upwards with all their force. The
weight of the baggage being thus partly supported,
the mule was able to proceed, and it was really

curious to see the gravity and caution with which the party regained the road.

During this singular operation, one of the party was for a long time endeavouring to catch his mule, who had escaped, and who managed just to keep out of his reach. When his master ran he ran: he followed his example when he walked, and at last, when my companion threw himself down on the snow quite exhausted, the cunning creature stood still and looked at him.

As I found that my mule still went very well, I cut across the snow, and saved more than a mile, though I had some places to descend which no animal but a mule could have accomplished. The melting of the snow had in some places undermined it, and as I travelled over the surface I could hear a torrent rushing under the feet of the mule. Several times I got off to walk, but was obliged to remount, as these animals will not be led by the bridle. My mule was getting tired, his back was rather sore, and so were his feet, when I came to a stream of water about a foot broad, but deep, and which was running under the snow we were crossing. The snow had fallen into this stream in two

or three places, both above and below me, and I
was quite sure it would not bear; so, in order that
the mule should tumble by himself, I rode to the
very edge, and then dismounting, put the bridle
over his neck, and crossing the little stream, I
endeavoured to persuade him to follow me, but he
would not think of it; it was but one step, yet he
would not make it.

I then resolved to back him over it, and accordingly taking hold of the Mameluke bit which was
in his mouth, I tried to turn him round. He
would open his mouth, and allow his head to come
round to his shoulder, but he knew what I wanted,
and nothing could persuade him to move his legs.

I could bear it no longer, so without a witness
but the wild mountains about me, I beat him on
his nose; however it was no use, he would not move,
and he looked so placid that I could not long be
angry with him, and therefore I gave the point up
and mounted him. The moment I was on his
back, he walked on; as I expected, the snow broke
in, and down he fell upon his nose; however, he
floundered through it, and then continued as patient as if nothing had happened, sometimes prick-

ing up his ears and looking at his path, as if some great curiosity, or some great danger, was before him, and then stopping to bray after his companions, during which nothing would induce him to proceed.

In about an hour we got out of the snow, and then continually descending, the country soon began to assume a different appearance; and when we afterwards came to the first trees, we fancied that we were beholding a most beautiful country, and our whole party were making repeated observations on the particular charms of the scenery, and were pointing out spots which they agreed would be the most delightful situations for villages and cottages.

In returning from several expeditions which we had before made to mountains, to inspect mines, I had always observed how very beautiful the plains looked after a short absence from vegetation, and I endeavoured to keep the observation in mind in viewing the scenes before me. Yet, upon the most deliberate reflection, I was of opinion that the climate was lovely, and that although the ground was rocky, the trees had a verdure and a luxuriance that I could not sufficiently admire; but when we returned over these same spots, after living in Chili, we all acknowledged the erroneous opinions we had formed, and were surprised to find the climate severe, the country bleak, and vegetation stunted by the continual frosts and violent winds.

I was now joined by two of my party, and we proceeded along a stream whose course guided us as on the other side. The torrent, however, was much more rapid, and it was very pleasing to see it rushing in a contrary direction to that which we had so long followed. We were riding close to a very high perpendicular mountain which was on our right, and were all looking up it, and making remarks upon its singular formation, when we heard a sound like the sudden explosion of a mine, and a large piece of the rock was instantly seen falling. The sound was exactly like that described, but I should think it must have proceeded from the rock having struck against some part of the cliff; however one of the party exclaimed " Oh! it is all coming!" and off he darted.

The other and I stood still, and we were much amused with the appearance of the fugitive, who bending over his mule, as if the mountain had al-

ready been on his shoulders, was kicking and spurring and beating his mule, and in this attitude actually rode out of our sight, without once turning
to look behind him.

When we came up to him, " What, did you not
see," said he, " the whole face of the mountain
moving, and smoke piping out of all the crevices?"
He added he had heard that Chili was full of vol-
canoes, that he considered the whole mountain was
coming upon him, and that therefore he certainly
did ride for his very life.

As our mules were very tired with the fatigue
they had undergone in climbing the Cumbre, we
stopped earlier than usual, at an uninhabited house
called La Guardia, where there was some food for
the mules, but as the house was full of fleas, most
of us slept on the ground outside.   A little after
midnight, as soon as the moon was up, we again
mounted our mules, but as the capataz was very
slow in loading the cargas, I rode on with one of
the party.

We came to several torrents and laderas, and the
former in the dark were passed very unwillingly,
for, as my companion very justly said, " If one is to

be carried away, one would like to see where one is going." As soon as the sun was up, we found it oppressively hot; and as our mules were getting lame, we could only trot very gently. The country down which we descended was similar to that which has already been described, and we continued our course till we came in sight of the town of La Villa Nueva de los Andes, whose name explains that it is a new town built in the Andes.

It is situated on ground comparatively flat, but is surrounded by mountains, or rather, hills; for the features of the country are here on a smaller scale.

The town, like all towns in Chili, is built on the usual plan. The streets are broad, and at right angles, and they are consequently parallel or perpendicular to each other. In the centre of the town there is a Plaza or great square, on one side of which is a rude sort of abode called the Governor's house, where a number of dirty-looking soldiers without shoes, and with little on them but a poncho, are seen sitting under a corredor, or lying about asleep.

I rode up to the guard, and asked a man who had an old sword in his hand, where La Fonda (the inn) was. He settled the point very quickly by saying, "Fonda no hai;" however, I learnt that there was a house where travellers were occasionally received, and he directed me to it. When I got there, I found it locked up. I knocked at the door for some time in vain; at last, a woman from the opposite side of the street told me that the people were gone away, and that the house was empty.

It was summer, and the sun, which in Chili is always burning, was to us who had come down from the snow so exceedingly overpowering, that I found it necessary to get into the shade somewhere or other, so I told my story to the women, and asked them where we could get shelter, a dinner, or even anything to drink. They said that the woman at the corner pulperia (shop) sold lemonade; but, as I was setting off, I saw at a little distance a quantity of rich clover-grass which had just been cut, so I filled my arms with it, and walked towards my mule. The grass was delightfully green, and the smell quite refreshing. The mule pricked up his long ears as he saw me coming; I threw it down before him, and took the iron mameluke-bit out of his mouth. After eating some mouthfuls of

N

it, he began to look about him, and I have seldom
felt more provoked, than I was to see him walk
away from it, and in preference begin to eat some
hot, dry, dirty straw, which was lying on a dung-
heap.

We then went to the shop, and I asked the old
woman what in the world we were to do?—that we
had come out of the Andes, were going next morning to Santiago, or, as they term it, to Chili, and
that we wanted food and lodging for the night.
She told me that the only thing to be done, was to
hire a room, and then get a person to cook whatever we wanted.

This sounded hopeless, but I soon found that we
had no alternative; so leaving my companion to
drink a glass of lemonade, and to take a siesta in
the old woman's bed, I went out on foot, following
a little boy without shoes, and was at last led to
the door of one of the largest houses in the place.
The boy went inside, and in a short time he returned with a large key in his hand, followed by a
well-drest, elderly lady, who asked me to walk in.
I declined, and went with the boy some distance
down the street; at last he stopped at a door, un-

locked it, and we entered a room full of feathers and fleas, and without any glass in the window. " Aqui sta," said the boy, and he added that I was to pay two reals (ten pence) a day. He said I could get dinner cooked at the next house. I accordingly went there, and found a woman who had the remains of very great beauty, and her daughter, of about eighteen years of age, who very much resembled her.

They both received me with the greatest kindness, and they insisted on my lying down on the bed. The old lady asked me what I would have for dinner for my party, and I told her all we wanted was the very best dinner she could give us, and that I begged to leave it to her good taste and judgment.

Away she went to get all the " materiel," while her daughter attended to me. She brought me a plate of the most delicious cool figs I ever tasted, and then a glass of iced lemonade, and all the time I was eating the figs she was sitting by the bed-side pitying me.

In about two or three hours the party arrived, mules and men quite fagged and exhausted, and I

spoke to the capataz about starting early in the
morning.   He lived about two leagues from the
town, and by agreement was to provide us with
fresh mules for the baggage, and horses for ourselves; but I could see he was not inclined to be
off early, so I insisted on his bringing the mules
and horses that evening.   He said that they would
have nothing to eat, so I gave him two dollars to
buy grass, and off he went, promising that he
would be back in the evening.

I had just time to bathe, when our dinner was
ready; and as the young woman brought us dish
after dish, the party observed, first, that she was
the most interesting-looking girl they had ever seen,
and secondly, that they had never eaten a dinner
so well drest; but the same delirium which, on
coming from the snow of the Andes, had made
them " babble o' green fields," caused them to err
in their judgments on other parts of creation; and
really, when we returned from the plain to Villa
Nueva, our dinner was badly cooked, and the poor
young woman was only said to be " rather pretty!"

The evening arrived, but not the capataz or
his mules, and we did not know where to send for

him ; but an hour before day-break the peon came to say that the capataz had turned him away; that he had spent the two dollars I had given him in drinking with his wife ; that he had not given us the proper quantity of spare mules at Mendoza, and he begged us to take him before the governor.

The sun was already up, when the capataz arrived. He had brought several of the poor tired mules, fresh mules for the riders, and a broken-kneed horse for me; but he was himself mounted on a fine prancing horse. I took his horse from him, put my saddle upon it, and desiring my party to take him before the governor, galloped off towards Santiago.

The road soon became very bad, as the path ascends a cuesta, which it is necessary to climb and to descend by zig-zags; however, as soon as I got on level ground by myself I galloped along, and it was quite delightful to be thus reminded of the pace of the Pampas, after having crawled along so many days on the back of a mule.

I soon got to the house at which we had agreed to sleep, and which is about half way between Villa Nueva and Santiago. It is a pulperia (shop), and was filled with peons drinking; however, they had got bread and wine, and I sent a man off to get a sheep; there was also a nice stream of water for bathing. In the course of three or four hours, several of the party arrived on horses, and they were in high spirits at the triumph they had gained over the capataz. They said that the governor had heard their cause, and had then ordered them to give the capataz a hundred lashes, but that as they did not exactly know how or where they were to inflict this punishment, they begged him to have the goodness to change it; upon which the governor said, that if I preferred it, I might pay him only six dollars for each of his mules, instead of eight, which was the sum agreed for. The latter award was certainly the best of the two; and, accordingly, when the capataz arrived, I assured him that if he had behaved well I should have given him, in addition to his agreement, the usual " gratificacion;" but, for his cruelty to his mules, I should most certainly inflict upon him one of the punishments to which the governor had sentenced him; and I left him for some time, uncertain which of the two he was to receive.

We all slept in the yard of the pulperia, on the ground, and long before day-break we started. I galloped on by myself, and at first took the wrong path; but as soon as I found by my compass that it was leading me away from Santiago, I changed my course, and at last came to a fire, round which a family were sleeping. After the usual barking of the dogs was silenced, I was directed where to go, and I crossed a number of small hills, until I came to the large uncultivated plain of Santiago. I was more than two hours galloping across this plain, which, from want of irrigation, produces no sort of herbage, but only scattered shrubs.

When I got within two leagues of the city, I came to water, and then the road was occasionally a pontana (swamp), through which, not knowing the passes, I had great difficulty to wade. An English horse would certainly have stuck in it, but those of the country, being accustomed to it, walk through very slowly, extricating their legs with the greatest caution.

I was now met by, and I overtook, men, women, boys, priests, &c. on horseback, either coming from or going into town, all at a canter, and in very singular dresses. Many of the horses were carrying double, sometimes two giggling girls, sometimes a boy with his grandmother behind him; sometimes three children were cantering along upon one horse, and sometimes two elderly ladies; then a solitary priest with a broad-brimmed white hat and white serge petticoats tucked up all about him, his rosary dangling on his mule's neck, and his pale fat cheeks shaking from the trot. Milk, and strawberries, and water-melons, were all at a canter, and several people were carrying fish into the town tied to their stirrups. Their pace, however, was altogether inferior to that of the Pampas, and the canter, instead of the gallop, gave the scene a great appearance of indolence.

The spurs of the peons were bad, and their stirrups the most heavy, awkward things imaginable. They were cut out of solid wood, and were altogether different from the neat little triangle which just holds the great toe of the Gaucho of the Pampas.

On crossing the bridge, which is at the entrance of the town, the market was underneath me, on some low ground on the left. A number of people were selling fruit, vegetables, fish, &c., which were lying on the ground, and as the sun was now oppressively hot, each parcel was shaded by a small canvass blind, which was fixed perpendicularly into the ground.

As I rode along the streets I thought they looked very mean and dirty. Most of the houses had been cracked by earthquakes; the spires, crosses, and weathercocks, upon the tops of the churches and convents were tottering, and out of the perpendicular; and the very names of the streets, and the stories " Aqui se vende, &c.," which are over all the shops, were written as crooked and irregular as if they had been inscribed during an earthquake. They were generally begun with large letters, but the man had apparently got so eager about the subject, that he was often obliged to conclude in characters so small, that one could hardly read them, and in some places the author had thoughtlessly arrived at the end of his board before he had come to the end of his story.

The great Plaza (square) has a fountain in the middle, and the Director's palace on one side. This building looks dirty and insufficient; it is of a fantastic style of architecture, and its outline is singular rather than elegant: part of it is used as a guard-room. The soldiers were badly dressed; some were blacks, wearing gold ear-rings, some were brown, and some of a mongrel breed.

It was just eight o'clock as I rode across this square. The bell of one of the churches tolled, and every individual, whether on horseback or on foot, stopped; the men all pulled off their hats, the women knelt down, and several people called to me to stop. The guard at the palace presented arms, and then the soldiers crossed themselves; in about ten seconds we all proceeded on our respective ways. This ceremony is always repeated three times a day, at eight in the morning, at noon, and at eight in the evening. I inquired my way to the English hotel, and found there a hard-working, industrious Englishwoman, who was the landlady. She told me she had not " an inch" of room in her whole house, which was filled with what she termed " mining gentlemen." I asked her where I could go; she said she could not tell, but she offered to send one of her servants with me to a "North American lady," who sometimes took in strangers. I went accordingly, and was introduced into a room which had a mat, a few highly-varnished, tawdry, wooden chairs, and a huge overgrown piano-forte. One side of the room was glazed like a green-house, and looked into another small room. Two long, thin, vulgar-looking girls, who talked through their noses, now came in, and told me a long story about "mama," the object of which was, that mama was coming, and accordingly in she came. They were all at once asking me to be seated, and were inquiring into my history, when I informed the lady, that I had called to inquire whether she had accommodation in her house for strangers. "Oh yes, she had a very nice room which she could let to me; there was no bed in it, but she could lend me chairs." I asked to see it; to my horror and astonishment, she led me to the glazed side of her room, and opening the glass door, she told me, that was the room. I had a great deal of very troublesome business on my mind, and all I required for the very few days I was to be at Santiago, was a little quietness and solitude. "Good heavens!" said I to myself, as I looked out of this wretched lanthorn, "how could I wash or make myself at all comfortable, either in body or mind, in such a place as this? Those girls, and that terrible piano-forte, would be the death of me! I am afraid, madam," addressing myself to the old lady, "this will not exactly do," and then out of the room, and out of the house, I walked.

I went back to the Englishwoman, who was very civil. The sun was burning me to pieces, I was quite exhausted, and I begged her to let me lay down anywhere in the shade, for that I had ridden almost all night, and was tired. She replied that she had positively no place. I told her I had been sleeping on the ground for many months, and that she surely had some little corner in which I might go to sleep. She said, "Nothing but the carpenter's shop." "Oh!" I said, with delight, "that will do famously;" so she led me to the place, and in a few seconds I was fast asleep among the shavings.

In three or four hours my party arrived, and

the landlady had by this time hired two empty rooms for them, and afterwards one small one for me. She got me a table, with two chairs, and she told us we could breakfast anddine with all her guests. This was not a very agreeable arrangement, but furnished lodgings are not to be had at Santiago, and I had therefore no alternative that than of hiring an empty house, and then getting furniture and servants; but to clean the former, and break in the latter, were occupations which I had no wish to undertake, particularly as I was going so shortly to inspect mines in different directions.

I had several letters, which at Buenos Aires I had been requested to take to Santiago, and these I at once delivered to a person to whom I was addressed. I had a drawing very carefully rolled up and sealed, which I had been told at Buenos Aires was the picture of a child in England, for his mother at Santiago. The lady happened to live close to the house to which I had taken my letters; and as I thought the picture of her child would be very acceptable, I called and delivered it to her myself. She was in one of the best houses in the town, and was surrounded by a very nice family of all ages. While I was talking to her she opened and unrolled the paper, and after glancing at it for a moment, she passed it to her family, who looked at it one after another with an apathy which quite provoked me. It was then handed to me, and I no sooner saw what it was, than I bowed to the family, and left in the hands of the lady, not a picture of her child, but a school-boy's large, coarse chalk-drawing of the head of John the Baptist!

During the short time I was at Santiago, I was constantly occupied in gaining information, without which I could not have commenced my inspection of the mines; and as many unforeseen difficulties were impeding my progress, and occupying my attention, I had neither time nor inclination to enter into any sort of society, or to see any more of Santiago than what chanced to be going on in the streets.

The town is full of priests—the people are consequently indolent and immoral; and I certainly never saw more sad examples of the effects of bad education, or a state of society more deplorable. The streets are crowded with a set of lazy, indolent, bloated monks and priests, with their heads shaved in different ways*, wearing enormous flat hats, and dressed, some in white serge cowls and gowns, and others in black. The men all touch their hats to these drones, who are also to be seen in the houses, leaning over the backs of their chairs, and talking to women who are evidently of the most abandoned class of society. The number of people of this description at Santiago is quite extraordinary. The lower rooms of the most reputable houses are invariably let to them, and it is really shocking beyond description to see them sitting at their doors, with a candle in

---

* I was one day in a hair-dresser's shop at Santiago, when a priest came in to have his head shaved, and I stopped to see the operation. The priest was a sleek fat man of about forty, with a remarkably short nose and a sallow complexion. The man lathered him with the greatest respect, and then shaved the lower part of his head about an inch above his ears all round, and discovered bumps which a student of Gall and Spurzheim would have been shocked at. His head was as deadly white as young pork; and while the barber was turning the priest's head in different directions, I really thought it altogether the most uncivilized operation I had ever witnessed; and when it was finished, and the man stood up, he looked so very grotesque that I could scarcely refrain from laughing.

the back part of the room burning before sacred
pictures and images.

The power of the priests has diminished very
much since the Revolution. They are not respected ; they have almost all families, and lead
most disreputable lives. Still the hold they have
upon society is quite surprising. The common
people laugh at their immorality, yet they go to
them for images and pictures, and they send their
wives and daughters to confess to them. Three
times a day the people in the streets take off their
hats, or fall down on their knees. Every quarter
of an hour during the night the watchman of each
street sings as loud as he is able a prayer of " Ave
Maria purissima," and then chants the hour and
a description of the night.

During the day one constantly meets a calesh
drawn by two mules, driven by a dirty boy in a
poncho, and followed by a line of inhabitants with
their hats off, each carrying a lighted candle in
a lantern : every individual in the streets kneels,
and those who have windows towards the streets
(who are generally the females I have described)
are obliged to appear with a lighted candle. In the

inside of the carriage sits a priest, with his hands uplifted and clasped. In this system of depravity the great sinner pardons the little one. Sins are put into one scale and money into the other, and intent upon the balance, both parties forget the beauty and simplicity of the religion which they nominally profess.

The siesta at Santiago is as long as it is at Mendoza. The shops are shut at noon, and remain closed for four or five hours, during which time all business is at an end.

The climate of Santiago is similar to that of all the parts of Chili which I visited. The day in summer is burning hot; the nights delightfully cool. During the day, the sun, reflected from the mountains which surround the town on every side, and which, of course, obstruct the breeze, has a greater heat than is natural to the latitude. At night the cold air rolls down the snowy sides of the Andes, and fills the Chilian valleys with a cool atmosphere, which is unknown to the great plains on the other side of the Cordillera. The effect of this stream of cold air is very agreeable, and people, whose occupations screen them from the sun in the

day, enjoy their evening's ramble; and as the sky is
very clear, the climate of Chili is often described as
being extremely healthy. Yet the least learned,
but perhaps the most satisfactory proof of the
healthiness of a climate is not the brightness of the
stars, or the colour of the moon, but the appearance
of men's and women's faces; and certainly the people
of Chili in general, and of Santiago in particular,
have not a healthy appearance. The English there,
also, look very pale and exhausted, and although
they keep each other in countenance, it appeared to
me, that a strong dose of British wind, with snow
and rain, and a few of what the Scotch call " sour
mornings," would do them a great deal of good.

*   *   *   *   *

CONVENT at Santiago.—Group of people on the outside whispering and speaking through the keyhole, the hinges and the cracks of the door—turnabout filled with old linen—door half opened by a janitress to take in two large models on wheels, the one of a brown cow, the other of a brown bull—door of the chapel open—chapel divided into two parts by a double grating, one of iron, the other of wood ; the lattices about the size of those in a cottage window. At one end the altar glittering with silver, mummery, and candles ; at the other side of the grating the nuns assembled at vespers—some were sitting at the sides and back of the chapel —others kneeling in the middle, even close to the grating, and with their faces towards the altar. They appeared to be almost all very old, fat women, short and thick—complexions stained with garlic and oil, and countenances soured by long confinement. They were praying as if they were sick and tired of it, and as if they neither cared nor knew what they were saying. Four or five were playing on fiddles, which they held up to their necks like men—one was sawing an immense double bass, and another was blowing with a large hand-bellows into the lungs of a little organ, on which a sister-nun was playing. They all sang together, and I never heard sounds less melodious. Age had taken all softness from their voices, and had left nothing but a noise which was harsh, squeaking, and discordant. The women were old and ugly, and the scene altogether was saddening. Their dresses consisted of white caps and large black gowns— their hair was concealed, and their features were so hard, that it was difficult to say whether they were old men or old women:—the serge gown concealed their figures—figures which were intended as the ornaments of creation. When one fancied the lives they might have led—the assistance they might have afforded to society—the friendships they might have enjoyed, and the pleasing natural duties they might have performed, it was melancholy to see them lost to the world, and only occupied in scream-

ing in Latin through iron bars to candles and
pictures.

On my right there was a young monk, who
remained on a bench close to the wall all the time I
was there.   He was confessing a nun through some
holes in a plate of tin, which was let into the
convent wall which separated them ; and since the
days of Pyramus and Thisbe, there can never have
been a more regular flirtation.   The monk was
much more anxious to talk than to hear, and I
could not help smiling when I saw him with great
eagerness of countenance putting sometimes his
mouth, and sometimes his ear, to the tin plate.
However, when I turned towards the group of old
nuns who were before me, I felt that it mattered
but little to society, whether they were confessing
their old sins, or planning new ones ; but it was
distressing to think that the young and the innocent,
who were rising in the world, were still the victims
of such a mistaken custom—for surely nothing can
tend to blunt the good feelings of the young more
than the reflection that even their thoughts of yes-
terday are already recorded by a man ; and if an
evil genius wished to prepare a man who should be

peculiarly unfitted for so delicate a confidence, what
could he do better than doom him to idleness and
celibacy, deny him children of his own, and feed
him upon oil and garlic?

*   *   *   *   *

JOURNEY TO THE GOLD MINE OF EL BRONCE DE PETORCA.

At about two o'clock in the morning we got up, and before we had eaten our breakfasts, the mules arrived with two peons. There were two mules for each person, and they were all driven loose into the yard. " Come now ! Vamos !" said one of the Cornish miners, who was always cheerful and ready to start, upon which the party all got their bridles and went down into the yard. The capataz took my bridle and promised to give me a good beast, and I stood for a few moments looking down upon the group from the large corredor or balcony. Each man was choosing his own mule; and as, from sad experience, he had learnt the difference between riding a good mule and a bad one, it was a point of some consequence. It was amusing to see each individual trying to look an animal in the face, to guess his character by the light of the moon, while the cunning creature, aware of his intention, was constantly hiding his head among his comrades, and turning his heels towards every person who approached him. As soon as the mules were saddled, which was always a troublesome and dangerous operation, we mounted, and rode out of the yard followed by the loose mules, who trotted after the madrina, or bell-mare, which was driven on by one of the peons.

As we passed through the streets the watchmen were singing the hour, with the usual hymn of " Ave Maria purissima;" and it was quite singular to hear their different ways of chanting it.

Our road passed across the plain of Santiago, and although we cantered, it was nearly three hours before we got to the mountains, and then for the whole day we had either to climb up one side of a barren mountain, or to scramble down the other. These mountains, from want of rain, afford scarcely any pasture: the soil upon them is cracked in a most singular manner, and the fissures are so deep and frequent, that it is apparently dangerous to ride over them.

After travelling until our mules were quite tired, we arrived, after the sun had set, at a small hamlet of mud huts. There had been a church, but the great earthquake of 1822 had converted it into a heap of ruins. The scene in the village was a very gay one. It was Christmas, and the usual festivities were going on. There were two or three rooms built of boughs, and filled with young women and Gauchos, who were dancing to the music of a guitar. On our arrival we had been led to the hut of a man who was the richest in the village; and as soon as we had taken our saddles into his house, we went out to join the dance. The sight of a few unexpected strangers added to the cheerfulness of the scene; the guitar instantly sounded louder, and the people danced with greater vigour. Round the room were rough poles as benches, on which sat the ladies who had danced; their partners were seated on the ground at their feet, and their earnest attentions cannot exactly be described. We were received with great hospitality, and in two minutes I saw my party all happy, seated on the ground, and as completely *enfans de famille*, as if they had been born there.

After remaining with them a short time, I returned to the hut. I found the master very sulky;

he had turned all our saddles out of his house, and for some little time he would not speak to me; however, I insisted that he should point with his finger where the saddles were, and accordingly I found them on the ground, outside a little hut, in which was one of the miners cooking our supper: however, we had slept so long in the open air, that it was of little consequence. I must do this man the justice to say, that though he was naturally a sulky fellow, he had intended to do right. He wished to have done the honours of his hut to strangers, and he accordingly gave the Cornish miner some eggs, but the man intending to pay for them, honestly told him there were not half enough, which the landlord considered as a breach of politeness.

While I was sitting on a horse's head, writing by the blaze of the fire, I saw two girls dressing for the ball. They were standing near a stream of water, which was running at the back of the hut. After washing their faces, they put on their gowns, and then twisting up their hair in a very simple pretty way, they picked, by the light of the moon, some yellow flowers which were growing near them.

These they put fresh into their hair, and when this simple toilette was completed, they looked as interesting, and as nicely dressed, as if " the carriage was to have called for them at eleven o'clock ;" and in a few minutes, when I returned to the ball, I was happy to see them each with a partner.

In the morning, before day, we started, and for many a league my companions were riding toge- ther, and discussing the merits of their partners. The country we rode over was mountainous, and it was very fatiguing both to mules and riders. I had just climbed up a very steep part of the mountain, and, with one of my party, was winding my mule through some stunted trees, when I sud- denly met a large-headed young man, of about eighteen years of age, riding his horse at a walk, and with tears running, one after another, down his face. I stopped, and asked him what was the mat- ter, but he made no reply. I then asked him how many leagues it was to Petorca, but he continued crying, and at last he said, " He had lost. . . . ." " Who have you lost ?" said I, debating whether it was his mother or his mistress. The fellow burst into a flood of tears, and said, " Mis espuelas" (my spurs), and on he proceeded.   One cannot say much for the lad's fortitude, yet the loss of spurs to a Gaucho is a very serious misfortune.   They are in fact his only property—the wings upon which he flies for food or amusement.

The sun was getting low, and the mules quite tired with the rocky barren path on which they had toiled, when we came to the top of a mountain, from which we suddenly looked down upon the valley of Aconcagua, which is a long narrow plain, irrigated by a fine stream of water.   The contrast was quite extraordinary ;—the colour of the trees and grass was black rather than green, and vegetation so rank and luxuriant, that the huts literally appeared smothered in the crops around them.   This picture is one which is constantly met with in Chili ; and as the produce of these plains, when irrigated, is greater than that of any of the world, Chili has often been called one of the richest countries.   But although these productive spots deservedly attracted the attention of the Spaniards, who found that the necessaries of life were there so easily obtained, yet the country is generally so mountainous, and so large a proportion of it is incapable of irrigation, that its population must hereafter be infinitely less than that of the Pampas, although at present it very much exceeds it.

On getting into the small town of Aconcagua, the church of which is in ruins, and almost every house cracked by earthquakes, we found the same sort of festivities which we had joined the evening before, but they were less interesting, because they were more formal. The Plaza (square) was covered with sheds, in which were people dancing, and when we rode up to the fonda, or inn, we saw the yard filled with people, sitting in bowers made of branches of trees, with others dancing or drinking.

We were eating our dinner at a small table in the yard, when a person came up and offered us a room at his house, and in the evening he came to take us to it. When he unlocked the door, which was on the ground-floor, we found the room filled with sacks of Indian corn, hides, rubbish of all sorts, and swarming with fleas; however, we made room, and slept there, and in the morning, after thanking the man for his lodging, we breakfasted

at the fonda, where we might have slept much
better.

Early the next morning we started on our fresh
horses and mules, leaving the tired ones in a potrero, or field, and visited a silver-mine, which was
within a league of the town. We then pursued
our course over barren mountains, and at about
twelve o'clock in the day we reached the village
of Petorca, which consists of one long principal
street, with other short ones at right angles.
The church, like that at Aconcagua, was overturned by the earthquake of 1822, and the walls
of the houses were cracked and rent from top
to bottom.

I had a letter of introduction to the principal
person, who was extremely polite, and was very
anxious that we should spend the evening with
him ; however, I at last prevailed upon him to get
us fresh mules, and about two o'clock, after we had
nad some refreshment, we set off with him to visit
some trapiches and mills which had existed before
the earthquake. We found the roofs shaken from
two of the huts, and the rest tottering. The two
mills were so completely annihilated, that it was

difficult to trace the foundation on which they had stood, and the water was diverted from its course.

In the evening our landlord gave us a most excellent supper, and the following morning, an hour before sunrise, we started to inspect the gold mines of El Bronce de Petorca, which were six miles from the village, and about a hundred and sixty from Santiago.

I visited this mine accompanied by a very intelligent Chilian miner, who with several of his comrades was in a mine on this lode a hundred fathoms deep, when the great earthquake of the 19th of November 1822, which almost destroyed Valparaiso, took place. He told me that several of his comrades were killed, and that nothing could equal the horror of their situation. He said that the mountain shook so that he could scarcely ascend; large pieces of the lode were falling down, and every instant they expected the walls of the lode would come together, and either crush them or shut them up in a prison from which no human power could liberate them. He added, that when he got to the mouth of the mine the scene was very little better: there was such a dust that he could not see his hand before him; large masses of rock were rolling down the side of the mountain on which he stood, and he heard them coming and rushing past him without being able to see how to avoid them, and he therefore stood his ground, afraid to move. In almost all the mines which we visited in Chili we witnessed the awful effects of these earthquakes, and it was astonishing to observe how severely the mountains had been shaken.

We got back to Petorca by ten o'clock, and as our host said he could give us fresh mules, I sent ours quietly on, and we agreed to start as soon as we had had a couple of hours' sleep.

After taking leave of our kind host, and bowing to the ladies, who were all standing at their doors, I went to the mule which had been provided for me, and saw by the wrinkles in his nose that he had some mischief in his head : however, he stood perfectly still, and allowed me to put my foot into the stirrup; but as soon as I threw my leg over him he jumped sideways about a yard; my heel went on to the top of some baggage which was upon the back of another mule, and my long Gaucho's spur got entangled in it. The mule, seeing that his plot had succeeded, began to kick, and with one leg up in the air, it was quite impossible to keep my seat. I fell on my head, and was stunned by the fall: however, as soon as I recovered I remounted him, expecting that he would kick again—*au contraire*, he was perfectly satisfied with what he had done, and he proceeded as quietly as a lamb.

* * * * *

GOLD MINE OF CAREN.

AFTER inspecting the old holes which had been worked on the lode, and gazing with great interest at the Pacific, which was apparently hanging in the air beneath us, we descended the side of the rock, sometimes upon hands and knees, for about three hundred and fifty feet, until we came to the hut where we had slept. The situation of this hut was singularly perilous. The path which ascended to it from the plain was so steep, that in riding up we constantly expected to tumble backwards over the tails of our mules; and when we got near the hut, the muleteers declared that it was altogether impossible to proceed, and this was so evident, that we dismounted and scrambled over the loose stones until we got to the hut.

The mine had not been worked for a hundred years, but it was now for sale. The hut had been just built, and a couple of miners ordered to live in it. A small space had been scarped out for the

foundation of the hut, which was so close to the precipice that there was not room to walk round it.
Above it, on the mountain, were loose rocks, which
by the first earthquake would probably be precipi-
tated. Beneath was the valley, but at such a depth
that objects in it were imperfectly distinguished. I
consulted with the two mining Captains, and we all
agreed that the plain was about three thousand feet
beneath us ; but this only gives our imperfect idea
of it, and is probably altogether incorrect ; for
although I spent some months among the Andes, I
was always deceived in the distances, and found
that my eye was altogether unable to estimate pro-
portions to which it had never been accustomed—
a trifling but a very striking proof of which occurred at this hut.

We were sitting with the native miners, when one
of my men called out that there was a condor, and
we all instantly ran out. He had been attracted
by the smell of a dead lamb, which we had brought
with us, and which was placed upon the roof of
the hut. The enormous bird, with the feathers of
his wings stretched out like radii or fingers, ma-
jestically descended without the least fear, until

apparently he was only ten or fifteen yards above
us. One of the men fired at him with a gun loaded
with large shot—his legs fell, and he evidently had
received the whole of the charge in his chest; yet
he instantly bent his course towards the snowy
mountains which were opposite to us, and boldly
attempted to cross the valley; but, after flying for
many seconds, he could go no further, and he began
to tower. He rose perpendicularly to a great
height, and then, suddenly dying in the air—so that
we really saw his last convulsive struggle—he fell
like a stone.

To my astonishment, he struck the side of the
mountain apparently close to us; and as I looked
at him lying on the rock, I could not account for
his being so very near us, (apparently thirty or
forty yards,) for as he had evidently fallen perpendicularly, the distance which separated us was of
course the hypothenuse of a right angled triangle,
the base of which it had taken him many seconds
to fly.

I sent one of the Chili miners, who were accustomed to descend the mountain, to fetch him, and I
went into the hut, and remained eight or ten mi-

nutes. On coming out, and asking for the bird,
I was surprised to see that the man was not half-
way to him; and although he descended and
ascended very actively, his return was equally long.
The fact was, that the bird had reached the
ground a great distance from us; but this distance
was so small in proportion to the stupendous objects around us, that, accustomed to their dimensions, we were unable to appreciate it.

JOURNEY TO THE SILVER MINE OF SAN PEDRO NOLASCO.

As soon as we returned to Santiago from the gold mine of Caren, we ordered fresh mules; and the next morning, before day-break, we set off to inspect the silver mine of San Pedro Nolasco, which is in the Andes, about seventy-five miles south-west from Santiago. For a few miles we traversed the plain of Santiago, which was cool and refreshed by the night air: just as the day was dawning we reached the foot of the mountains, and then following the course of a large rapid torrent, we continued for several hours on the east side of it, climbing along a path which appeared to overhang the water.

As the sun gradually rose, the mountains on the opposite side were scorched by the heat, while we for several hours were in the shade and cool; but the line of shadow, after crossing the torrent, gradually approached us, the sun at last looked over the high

mountains which were above us, and that instant
commenced the fatigue of the day.

The valley of Maypo, down which the stream
descended, is one celebrated in Chili for its beauty.
Bounded on both sides by the barren mountains of
the Cordillera, this delightful vale winds its course
on both sides of the river or torrent of Maypo;
and although it is uncultivated, yet it is orna-
mented with a great variety of shrubs and fruit-
trees.

For several leagues we passed trees loaded with
ripe cherries, and peach trees which were bending
towards the ground with the weight of their crop.
The ground underneath was covered with the
peach-stones of the last year's produce, and there
must be thousands of these trees whose fruit has
never once been tasted by man. The ground, although it produced shrubs and trees, had no appearance of pasture, which cannot in a hot climate
exist without irrigation.

After travelling about thirty miles, we crossed
the torrent of Maypo, on a suspension-bridge of
hide ropes, the construction of which I examined
with great attention, as I was surprised to find it

exactly similar to those which I have seen con-
structed in England of iron, although this bridge
has been there beyond the memory of man. The
path across it was covered with hurdles, and as the
torrent was much swollen, the water was rushing
over it with great velocity, which, of course, made
the bridge incline very much. Our mules were unwilling to cross it, and I certainly should have thought
it dangerous, had not a man who was on the opposite side beckoned to us to come over. The bridge
bent with the weight of the mules, and the water
rushed with great violence against them, but they
leaned against it, and we all passed it without
accident; and in returning rode over it in the
dark.

After continuing our journey about four miles,
we came to a small establishment for reducing the
ores raised from San Pedro Nolasco, and for the interesting process of amalgamation, and we remained
here for the evening to inspect it.

Without entering into a description of the establishment, it will only be observed, that the works
were laid out with a great deal of ingenuity, with a
very happy regard to economy, and that, although

they of course did not possess many of the me-
chanical advantages which a large capital might
have afforded them, yet they were on a plan suited
to the resources of the country, and upon the
whole were well adapted for the economical
reduction and amalgamation of ores upon a small
scale.

The next morning, before sunrise, we continued
our course towards San Pedro Nolasco, and for
four or five hours followed the course of the river.
The valley became narrower, and as we proceeded
the trees and shrubs became smaller and more
stunted—around us on every side were the Andes
covered with snow. Our path was in many places
very dangerous, being infinitely more so than any
of the parts we had crossed in coming from Mendoza over the Cordillera. The laderas were literally only a few inches wide, and were covered
with stones, which were so loose, that every instant
they rolled from under the mules' feet, and fell with
an accelerating violence into the torrent. As I
rode almost the whole of the day by myself, I
would willingly have got off; but the mules will
never lead, and besides this, when once a person is

on the ladera, on the back of his mule, it is impossible to dismount, for there is no room to get off, and the attempt to do so might throw the mule off his balance and precipitate him into the torrent, which was at an extraordinary depth beneath. In some few places, the path was actually washed away, and the mule had only to hurry over the inclined surface the best way he could; but the manner in which these patient animals preserve their footing is quite extraordinary, and to know their value one must see them in the Cordillera. After passing two or three very violent torrents, which rushed from the mountains above us into the river beneath us, we came to one which looked worse than those which we had with great difficulty crossed; however, we had no alternative but to cross it, or return to Santiago. We attempted to drive the loose mules across, but one had scarcely put his feet into it, when he was carried away, and in less than twenty yards the box which he had on his back was dashed to pieces, and its contents were hurried down the surface of the stream. In order to get across, we put a lasso round our bodies, and then rode through; but the holes were so deep, that the water occasionally came over the neck of the mule, and we passed with great difficulty. These poor creatures are dreadfully afraid of crossing these torrents; it is only constant spurring that obliges them to attempt it, and sometimes in the middle of the stream they will refuse to advance for several seconds. When the water is very deep, the arrieros always tie the lasso round their bodies; but I never could conceive it was any security, because if the torrent will dash a wooden box to pieces, a man's skull would surely have a very bad chance. I was, therefore, always very glad when I found myself across them; and, as our lives were insured in London for a large sum of money, I used often to think, that if the insurers could have looked down upon us, the sight of the laderas and of these torrents would have given a quickness to their pulse, a flush to their cheek, and a singing in their ears, very unlike the symptoms of placid calculation.

Shortly after passing this torrent, we turned towards the south, and began to climb the mountain of San Pedro Nolasco, which I can only describe by saying, that it is the steepest ascent

which we ever made in all our expeditions among
the Andes. For five hours we were continually
holding on by the ears or neck of our mule, and
the path was in some places so steep, that for a
considerable time it was quite impossible to stop.
We soon passed the limits of vegetation. The
path went in zig-zags, although it was scarcely
perceptible, and if the mules above us had fallen,
they would certainly have rolled down upon us,
and carried us with them.

In mounting we constantly inquired of the
arriero, if the point above our heads was the summit, but as soon as we attained it, we found that
we had still higher to go. On both sides of us we
now came to groups of little wooden crosses,
which were the spots where people formerly employed in the mine had been overtaken by a storm,
and had perished. However, we continued our
course; and at last, gaining the summit, we found
ourselves close to the silver lode of San Pedro
Nolasco, which is situated on one of the loftiest
pinnacles of the Andes. A small solitary hut
was before us, and we were accosted by two or
three wretched-looking miners, whose pale coun-

tenances and exhausted frame seemed to assimilate with the scene around them. The view from the eminence on which we stood was magnificent—it was sublime; but it was, at the same time, so terrific, that one could hardly help shuddering.

Although it was midsummer, the snow where we stood was, according to the statement made to me by the agent of the mine, from twenty to a hundred-and-twenty feet deep, but blown by the wind into the most irregular forms, while in some places the black rock was visible. Beneath was the river and valley of Maypo, fed by a number of tributary streams, which we could see descending like small silver threads down the different ravines. We appeared to have a bird's-eye view of the great chain of the Andes, and we looked down upon a series of pinnacles of indescribable shapes and forms, all covered with an eternal snow. The whole scene around us in every direction was devoid of vegetation, and was a picture of desolation, on a scale of magnificence which made it peculiarly awful; and the knowledge that this vast mass of snow, so cheerless in appearance, was created for the use, and comfort, and happiness, and even luxury of man; that it was the inexhaustible reservoir from which the plains were supplied with water,—made us feel that there is no spot in creation which man should term barren, though there are many which Nature never intended for his residence. A large cloud of smoke was issuing from one of the pinnacles, which is the great volcano of San Francesco; and the silver lode, which was before us, seemed to run into the centre of the crater.

As it was in the middle of the summer, I could not help reflecting what a dreadful abode this must be in winter, and I inquired of our guide and of the miners concerning its climate in that season. They at first silently pointed to the crosses, which, in groups of three and two and four, were to be seen in every direction; and they then told me, that although the mine is altogether inaccessible for seven months in winter, yet that the miners used to be kept there all the year. They said that the cold was intense, but that what the miners most dreaded were the merciless temporales, or storms of snow, which came on so suddenly that many miners had been overtaken by them, and had perished when not a hundred and fifty yards from the hut. With these monuments before my eyes, it was really painful to consider what the feelings of those wretched creatures must have been when, groping about for their habitation, they found the violence of the storm unabating and irresistible. It was really melancholy to trace, or to fancy I could trace, by the different groups or crosses, the fate of the different individuals. Friends had huddled together, and had thus died on the road; others had strayed from the path, and from the scattered crosses, they had apparently died as they were searching for it. One group was really in a very singular situation; during a winter particularly severe, the miners' provisions, which consist of little else than hung-beef, were gradually failing, when a party volunteered, to save themselves and the rest, that they would endeavour to get over the snow into the valley of Maypo, and return if possible with food. They had scarcely left the hut, when a storm came on, and they perished. The crosses are exactly where the bodies were found; they were all off the road; two had died close together, one was about ten yards off, and one had climbed to the top of a large loose fragment of rock, evidently to look for the hut on the road. The view from San Pedro Nolasco, taking it all together, is certainly the most dreadful scene which in my life I have ever witnessed; and it appeared so little adapted or intended for a human residence, that when I commenced my inspection of the lode, and of the several mines, I could not help feeling that I was going against nature, and that no sentiment but that of avarice could approve of establishing a number of fellow-creatures in a spot, which was a subject of astonishment to me how it ever was discovered..

As the snow was in many places fifty feet deep on the lode, I could only walk on the surface from one bocca-mina to another; but when I had done this, I took off my clothes, and went down the mine which it was my particular object to inspect. All the rest had long ago been deserted, but in this one there were a few miners, lately sent there, who were carrying on the works on the old system which had been exercised by the Spaniards, and to which these men have all their lives been accustomed.

At first we descended by an inclined gallery or level, and then clambered down the notched sticks, which are used in all the mines in South America as ladders. After descending about two hundred and fifty feet, walking occasionally along levels where the snow and mud were above our ancles, we came to the place where the men were working. It was astonishing to see the strength with which they plied their weighty hammers, and the unremitted exertion with which they worked; and strange as it may appear, we all agreed that we had never seen Englishmen possess such strength, and work so hard. While the barreteros, or miners, were working the lode, the apires were carrying the ore upon their backs; and after we had made the necessary observations, and had collected proper specimens, we ascended, with several of these apires above and below us.

The fatigue of climbing up the notched sticks was so great, that we were almost exhausted, while the men behind us (with a long stick in one hand, in the cloven end of which there was a candle) were urging us not to stop them. The leading apire whistled whenever he came to certain spots, and then the whole party rested for a few seconds. It was really very interesting, in looking above and below, to see these poor creatures, each lighted by his candle, and climbing up the notched stick with such a load upon his back, though I occasionally was a little afraid lest one of those above me might tumble, in which case we should have all preceded him in his fall.

We were quite exhausted when we came to the mouth of the mine; one of my party almost fainted, and as the sun had long ago set, the air was so bleak and freezing—we were so heated—and the scene was so cheerless, that we were glad to hurry into the hut, and to sit upon the ground round a dish of meat, which had long been ready for us. We had some brandy and sugar, and we soon refreshed ourselves, and I then sent out for one of the apires with his load. I put it on the ground, and endeavoured to rise with it, but could not, and when two or three of my party put it on my shoulders I was barely able to walk under it. The English miner who was with us was one of the strongest men of all the Cornish party, yet he was scarcely able to walk with it, and two of our party

who attempted to support it were altogether unable, and exclaimed " that it would break their
backs."

The load which we tried was one of specimens
which I had paid the apire to bring up for me,
and which weighed more than usual, but not much,
and he had carried it up with me, and was above
me during the whole of the ascent.

While we were at one end of the hut, drinking
brandy-and-water, seated upon our saddles, and
lighted by a brown tallow-candle which was stuck
into a bottle, and which was not three yards from a
hide filled with gunpowder; the few miners we
had seen at work had been relieved by others who
were to work through the night.   They came into
the hut, and without taking the least notice of
us, prepared their supper, which was a very
simple operation.   The men took their candles out
of the cloven sticks, and in the cleft they put a
piece of dried beef; this they warmed for a few
seconds, over the embers which were burning on
the ground, and they then eat it, and afterwards
drank some melted snow-water out of a cow's-horn.

Their meal being over, they then enjoyed the

only blessing fortune had allotted to them, which was rest from their labour. They said nothing to each other; but as they sat upon the sheep-skin, which was the only bed they had, some fixed their eyes upon the embers, while others seemed to ruminate upon other objects.

I gave them what brandy I had, and asked them if they had no spirits, to which they gave me the usual answer, that miners are never allowed to have spirits, and with this law they seemed to be perfectly satisfied.

When one contrasted their situation with the independent life of the Gaucho, it was surprising that they should voluntarily continue a life of such hardship.

*   *   *   *   *

DEPARTURE FROM SANTIAGO.

* * * * *

DECEMBER 31st, Santiago, midnight.—Mules arrived for us to recross the Cordillera to return to Buenos Aires—a large drove—two mules for each person—spare ones for the baggage. At one o'clock in the morning the mules were laden and ready—went across the street to the fonda, to get some breakfast, which was laid for us at one end of a long table—at the other end were two Scotchmen sitting without their coats, waistcoats, or neckcloths—(midsummer.)

They had been drinking-in the new year—in their heads there was " mair brandy than brains," yet their hearts were still true to their " auld respected mither." The room was evidently moving round them—they were singing (with action) " Auld lang syne," and the one that was pitted with the small-pox seemed to feel it as much as the other they held ·out glasses to us, and begged us

to join them—we declined—amusing contrast between them and the gravity of my party, drinking
tea, with their pistols in their belts, and prepared
for a long journey—full chorus of Rule Britannia,
then God save the King ; shook hands with the two
Scotchmen—drank half a glass of their brandy,
and then mounting our mules—we groped along in
the dark towards the black mountains of the Cordillera.

*   *   *   *   *

RETURN TO MENDOZA.

*   *   *   *   *

Got to Uspallata late in the evening with two of the
party; at sunset the rest arrived.   Mules tired—
the maestro de posta had three horses, and being
anxious to get on to Mendoza (ninety miles), three
of us rode all night.   We had three times travelled
the road, and therefore went by ourselves.   About
half way we saw a fire on the ground, and by the

blaze we perceived some person near it—rode up
to light our cigars, called several times, but found
no one.   On arriving at the hut near Villa Vicencia
we mentioned the circumstance, and were told it
was probably an Englishman who had passed the
hut that day on foot!—that he had probably been
afraid of us, and had concealed himself, or had run
away.

Rested, and then got fresh horses at Villa Vicencia. The sun was most dreadfully hot. We galloped across the plain—forty-five miles—each at our
best pace—proceeded straggling, like the wounded
Curiatii. I got into Mendoza three hours before
the second—he got in two hours before the third,
whose horse was tired on the road.

In riding along the plain I passed a dead horse,
about which were forty or fifty condors; many of
them were gorged and unable to fly; several were
standing on the ground devouring the carcass—the
rest hovering above it. I rode within twenty yards
of them: one of the largest of the birds was standing with one foot on the ground and the other on
the horse's body—display of muscular strength as
he lifted the flesh and tore off great pieces, some-

times shaking his head and pulling with his beak,
and sometimes pushing with his leg.

Got to Mendoza, and went to bed. Wakened by
one of my party who arrived : he told me, that
seeing the condors hovering in the air, and knowing
that several of them would be gorged *, he had also
ridden up to the dead horse, and that as one of
these enormous birds flew about fifty yards off, and
was unable to go any farther, he rode up to him,
and then, jumping off his horse, seized him by the
neck. The contest was extraordinary, and the
rencontre unexpected. No two animals can well be
imagined less likely to meet than a Cornish miner
and a condor, and few could have calculated, a year
ago, when the one was hovering high above the
snowy pinnacles of the Cordillera, and the other

---

* The manner in which the Gauchos catch these birds is to
kill a horse and skin him ; and they say that, although not a
condor is to be seen, the smell instantly attracts them. When
I was at one of the mines in Chili, I idly mentioned to a person
that I should like to have a condor : some days afterwards a
Gaucho arrived at Santiago from this person with three large
ones. They had all been caught in this manner, and had been
hung over a horse ; two had died of galloping, but the other
was alive. I gave the Gaucho a dollar, who immediately left
me to consider what I could do with three such enormous birds.

many fathoms beneath the surface of the ground in Cornwall, that they would ever meet to wrestle and " hug " upon the wide desert plain of Villa-Vicencia.	My companion said he had never had such a battle in his life; that he put his knee upon the bird's breast, and tried with all his strength to twist his neck; but that the condor, objecting to this, struggled violently, and that also, as several others were flying over his head, he expected they would attack him.	He said, that at last he succeeded in killing his antagonist, and with great pride he shewed me the large feathers from his wings; but when the third horseman came in, he told us he had found the condor in the path, but not quite dead.

THE PAMPAS.

I WAITED some time at the post-hut, talking with the old lady, who was always very kind and glad to see me, and was also extremely clever and entertaining; I then mounted my horse, and, after galloping nearly an hour, I overtook the coach just as it had reached the banks of the river Desaguadero, which was unusually deep and rapid. There was nothing but a small bark, but we lost no time in filling it with the luggage, and then made preparations for dragging the carriage through the river. I took off my clothes, and throwing them into the boat, I tied a silk handkerchief round my neck, and put my watch there to keep it dry. I had my pistols in my right hand, and I then rode into the river. The horse was instantly out of his depth, but he swam over very well. Just as I had scrambled up the bank, a man, dressed in a dirty-looking poncho, who lived in a hut* about a hun-

* The miners were one morning very much amused at the sight of a man who was asleep upon the ground near this hut.

dred yards off, came up and asked to be paid for the boat; I told him I would pay him as soon as the coach was over, and I asked him to take care of my pistols for me, and he accordingly took them to his hut.

We then set to work to get the carriage over, which was a very curious operation. The bank to descend to the river was much steeper than 45°, and it was therefore necessary to fix a peon, with his horse and lasso to the back part of the carriage, to prevent its oversetting; we had also lassos fastened as guys on each side. Two or three peons fixed their lassos to the end of the pole, and one swam across the river with a long drag rope, to which eight or ten horses were affixed to assist in dragging the carriage. As soon as these arrangements were made, the carriage was lowered down the bank, but its weight was so great that it dragged after it the peon and horse fixed to retain it; and while our party also were hauling at the rope, it was curious to see them all dragged down the bank. As soon as the carriage came into the river, although the wheels and

His wife had just risen, but he was still snoring, with his head lying on a bullock's skull, which had an enormous pair of horns.

perch were unusually high, it was nearly filled with water. In this state the peons whose lassos were fixed to the end of the pole, with all the horses at the drag rope, dragged the carriage slowly along the bottom of the river: however, when it was about half-way across, it would come no farther, and the horses which were on the steep bank had little power to draw. The carriage remained in this hopeless and singular situation more than an hour, during which time we were occupied in altering the drag ropes, and arranging them more advantageously.

I found the sun so burning hot, that several times I swam about on my horse to cool myself, and then galloped on the opposite side of the river, and I cannot express the delightful feeling of freedom and independence which one enjoys in galloping without clothes on a horse without a saddle.

When the horses and peons were ready, they all started together, and at last the carriage began again to move; and the peons then spurring, flogging, and cheering their horses, it was dragged up the bank.

While they were putting the luggage into the

wet carriage, I dressed myself, and then rode up to the hut to pay the man for his boat. He demanded twelve dollars, which I knew was too much, and I therefore refused to give it. In a moment, he was in a violent passion; he addressed himself sometimes to me, and sometimes to some Gauchos who were sitting drinking; and he was approaching me with menacing gestures, when I took my pistols off his table, and before I placed them into my belt, I put the muzzle of one of them against his front tooth, and told him very quietly, that I would pay him what was proper, but that if he demanded more, I would only pay him with that pistol. In an instant, the man desired one of the Gauchos to saddle him a horse, in order to ride to the Governor of St. Luis, who he said was a relation of his, and he then told me that he was himself a judge. I laughed at him, and telling him that he was a bad judge in his own cause, I left him, and rode after the coach.

In about half an hour the fellow overtook me, and without speaking he galloped by me. He was dressed in his judicial robes; that is, he had on a coarse blue jacket, with scarlet cuffs and collar, and a long sabre. I now continued my course for the remainder of this post, which is fifty-one miles, changing my horse when I overtook the droves of loose horses which preceded the carriage.

This stage is really one of the most singular examples of South American travelling which I have witnessed. We started with seventy horses, which were driven before us at a gallop. These horses were all loose; and the country hot sand, covered with trees and brushwood. The trees are principally the Algarroba; they were about the size and shape of apple-trees, and were sufficiently high to hide the horses. This drove of wild loose animals was driven by a man and a boy, and it was quite surprising, as I galloped along the road, to see these fellows constantly darting across the path before me, in close pursuit of the horses, which were never to be seen in the road. In the plains of grass it is even wonderful to see how the troops of horses are driven on, but in a wood it is much more astonishing; and it is a beautiful display of horsemanship to see the Gauchos galloping at full speed among the trees, sometimes hanging over one side of their horse, and sometimes crouching upon his neck to avoid the branches of the trees. The carriage road is a space cleared of large trees, but it is often covered with bushes, which bend under the carriage in a most extraordinary manner.

I arrived at the post some hours before the carriage, and had supper ready by the time it arrived. This post is only one stage from St. Luis; the postmaster is the brother of the governor of the province, and he was at St. Luis when I arrived, but his capataz asked me, with great seriousness of countenance, whether I was the person who had galloped after the judge at the Desaguadero, in order to shoot him. He told me that the said Juez had just passed, and had taken a fresh horse to get to St. Luis before I arrived there. We slept that night at the post, or rather on the ground before it; and it was curious, in the morning, to see the different groups of people, who had also slept there, dressing themselves;—men, women, and children, were all sitting up as if just risen from the grave— some were scratching themselves, some were rubbing their eyes, some putting on their hide sandals;— the hens were pecking about them, particularly round the table at which we had supped. The

great dogs, who had also just awakened, were
walking very slowly with their tails between their
legs towards the corral, where there is always a
supply of food for them. The infants were still
sleeping, each upon a lamb's skin, on the ground,
without a pillow, covered only with a piece of dirty
blanket, and sometimes the hens would perch upon
them. As soon as the horses were caught we set
off, and I galloped into St. Luis, and got there an
hour before the carriage. I found the post as
usual; there was nothing to be had—no fruit,
though in the middle of summer, and no milk. The
people of the post-house told me, that the Juez had
arrived there last night, and it appeared that his
story had been much inflamed by his gallop. As
soon as the carriage arrived, the Juez and an ordenanza, or horse-soldier, came up to the post, and
told me that I was to come immediately to the
governor. I had a white linen jacket on, which
was really too dirty to go in, so I resolved to put
on a coat. On opening my portmanteau, out came
a quantity of water, and I found that it had been
filled in passing the Desaguadero—my coat was consequently dripping wet; however, I put it on, and

as I knew the way, I galloped towards the barracks followed by the Juez and the Ordenanza. I found the square filled by a set of most wretched-looking persons, who were assembled to be sent to Buenos Aires to fight against the Portuguese. There were about three-hundred of them, and the night before they had endeavoured to gain their liberty, and had tried to overpower their guard. They were covered with old ponchos, but had very little on besides; they seemed to have been badly fed, and were altogether the wildest-looking crew I ever beheld.

The governor was standing in the middle of the square, surrounded by a number of officers, and I dismounted and walked up to him. He began, very hastily, by telling me the Juez's story; however, I asked him if he would allow me to tell mine. I told him that it was so much my duty to respect governors and governments, that if I had known the man who was before us had been in his employment, I would have respected him, though his conduct did not deserve it; but that instead of wearing the clothes he now had on, he was dressed in a dirty poncho—was drinking aquadiente with the Gauchos, and that I had therefore no idea he

R

was a Judge. I explained the circumstances, and
the governor then told the man that he had asked
too much, and that I was to pay him three dollars
less than he had demanded. The governor offered,
very obligingly, to lend me the money as I had no
change ; he paid the man, who had not a word to
say, and who had his ride, one hundred and eighty
miles, for nothing. I then went into the governor's
room, and mentioned to him that our carriage
wanted a trifling repair, but that the blacksmith
had told me he could not work at it without permission from him, as he was employed in making
chains to take the three hundred recruits down to
Buenos Aires. The governor very politely sent
for the smith, and desired him to work for me for
three hours; after which I made my bow, and then
galloped to the post.

While the smith was repairing the carriage, I
looked again at the town of San Luis. Each house
in the town has a large garden, in which there is
nothing but what they cannot prevent from growing,
such as fig trees, vines, peach trees. The walls of
their gardens are often towards the streets, which
gives the place so little the appearance of a town,

that the first time I came to San Luis, I actually asked a man how far I was from El Pueblo; to which he replied, that I was in it. From twelve till four or five every day, the whole population of the town is asleep, and when the people awake, they have no other idea than that of satisfying their hunger, by eating the old dish, carne de vacca. Far from having any luxuries, they have not even what we term common necessaries; and it seems incredible that there should be no individual in the whole town, or indeed the province, who even professes to know any thing of medicine or surgery; and that there is no shop at which one can purchase the simplest medicines. If a person is ill, he dies or recovers as it may happen, but he has no assistance. If he dislocates or breaks a bone, his friends may regret the accident, but he has no help. The Gaucho, who lives in his little hut on the Pampas, must necessarily be without medical assistance, and it is interesting to see his young family living so completely under the sole protection of Providence; but for the capital of a province to continue in such a state, shows an indolence, which its peculiar situation can only excuse.

The post-house of San Luis is also in a state
which would scarcely be credited. It is in nothing
better than the post-huts of the Pampas; it has no
window, the door cannot be shut, and it is more
filthy than can well be described. It was late
before the carriage was ready; however as I was
anxious it should get on, it started with three
changes of horses, about an hour before sunset, to
go to the next post, which is thirty-six miles. I
rode by a different road, and it was settled that we
should all get on by moonlight; however, as soon
as the sun set the weather began to look wild, and
it became very cloudy and dark. I continued to
gallop until I could not see my hand before me,
and as I knew there were many holes and biscacheros, we then slackened to an ambling canter. It
is really very nervous, disagreeable work even to
canter over a strange country when it is quite dark;
however, I was anxious if possible to reach the post,
as it was the nearest hut we could get to. I was cantering along, expecting every moment to tumble
head over heels, when my horse suddenly struck his
chest against the back of the Gaucho's horse, which
was standing still. As soon as I found out what it

was I spoke to the man, but I received no answer;
I then called out, when he told me from some distance, that he was feeling with his hands for the
path—that he could not find it,—and that there
were so many holes that, as we had lost ourselves,
it would be dangerous to proceed. I accordingly
dismounted, and, unsaddling my horse, I had instantly my bed ready. I could see nothing, but
the Gaucho and I made our beds side by side, and
as soon as we lay down he tied the horses' bridles
round his own neck, and he then was asleep in a
moment.

The country we were in was much infested by
salteadores (robbers,) but as I was always well
armed I felt quite secure, and in a short time I was
also asleep. About midnight I was awakened by
the rolling of thunder, and, sitting up, I saw by
the occasional flashes of lightning that I was lying
on brown coarse grass, and that there were here
and there a few shrubs. Some large heavy drops
of rain began now to fall, and I made up my mind
that we were to have a drenching shower; however, it was useless to move, for I did not know
where to go, so I took the usual precaution, which

is to place the skin which, in dry weather, one lies on, over my head, and I then went to sleep. Before the day began to dawn I was awakened by the Gaucho, who told me the horses were lost. I told him very sulkily to go and look for them, and, with my head under the skin, I again dropped off to sleep. I was awoke by the heat of the sun, and jumping up found that it was above the horizon, and that it was late. I looked earnestly round me, but, except a few shrubs, there was nothing but " the wind blowing and the grass growing,"—in every direction was a vast expanse of plain. I began to think that the man had returned to San Luis, and I really did not know what I should do. The sun was oppressively hot, and I was standing in despair, gazing at the recado which had formed my bed, when I heard the distant notes of a Spanish song behind me, and turning round I saw the Gaucho galloping towards me, and driving my horse before him. In a few moments he came up: my horse was of course without a bridle ; the fellow had played me the old trick of hiding it, and declaring it was lost. However, I was glad to get my horse upon any terms, and I cut a piece of hide, which served to guide him, and we then galloped towards the post, from which we were distant about thirteen miles.

I there got some breakfast, while they were catching another horse for me. They had neither bread nor milk, but I got some water, a couple of eggs, and an old woman warmed some charque for me over the embers. I was surrounded by several women and girls, all three-quarters naked, who asked me if I could give them maté or sugar, " por remedio ?" As soon as my horse was saddled, I purchased the bridle of the Gaucho who had stolen mine, and then galloped on. The country, which from Mendoza is covered with wood, now changes to the long brown and yellow grass, which, excepting a few straggling trees, is the sole produce of the remainder of the province of San Luis, and of the two adjoining provinces of Cordova and Santa Fé. In the whole of this immense region there is not a weed to be seen. The coarse grass is its sole produce; and in the summer when it is high, it is beautiful to see the effect which the wind has in passing over this wild expanse of waving grass: the shades between the brown and yellow

are beautiful—the scene is placid beyond description—no habitation nor human being is to be seen,
unless occasionally the wild and picturesque outline
of the Gaucho on the horizon—his scarlet poncho
streaming horizontally behind him, his balls flying
round his head, and as he bends forward towards
his prey, his horse straining every nerve : before
him is the ostrich he is pursuing, the distance between them gradually diminishing—his neck stretched out, and striding over the ground in the most
magnificent style—but the latter is soon lost in the
distance, and the Gaucho's horse is often below the
horizon, while his head shews that the chase is not
yet decided. This pursuit is really attended with
considerable danger, for the ground is always undermined by the biscachos, and the Gaucho often falls
at full speed ; if he breaks a limb his horse probably gallops away, and there he is left in the long
grass, until one of his comrades or children come to
his assistance ; but if they are unsuccessful in their
search, he has nothing left but to look up to heaven,
and while he lives drive from his bed the wild
eagles, who are always ready to attack any fallen
animal. The country has no striking features, but

it possesses, like all the works of nature, ten thousand beauties. It has also the grandeur and magnificence of space, and I found that the oftener I crossed it, the more charms I discovered in it.

On approaching the huts, it is interesting to see the little Gauchos, who, brought up without wants, and taught to consider the heaven over their heads as a canopy under which they may all sleep, literally climb up the tails of the horses which they are unable otherwise to mount, and then sport and gallop after each other, while their father's stirrups are dangling below their naked feet. In the foreground of Nature, there is perhaps no figure so beautiful as that of a child who rides well, and the picturesque dress of the little Gauchos adds very much to their appearance. I have often admired them as they have been sent with me from one post to another. Although the shape of their body is concealed by the poncho, yet the manner in which it partakes of the motion of the horse is particularly elegant. It is interesting, too, to see the heedless, careless way in which these little chubby-faced creatures ride, and how thoughtlessly they drive

their horses among biscacheros, which would break
in with the weight of a man.

When I got to El Morro I resolved to wait there
for the carriage, for I had the keys of my portmanteau, and both I and my party wanted money. El
Morro consists of a few mud huts, as usual without
windows; and as I stood at the door of the post-
room no human being was to be seen, except occasionally a woman with her hand or poncho shading
off the sun from her head as she crossed the broad
irregular street which divided the huts from each
other: here and there a horse was seen tied to the
outside of a hut, and a little tame ostrich was before the door running after flies: the atmosphere
was quivering with the heat, and resounding with
the shrill cry of millions of flies enjoying the sun.
There were no trees to be seen, and neither fruit
nor flowers to be had. I went to the woman of
the post to ask what she had got to eat: "Nada
(nothing), Senor," she replied. I asked for several things which, from seeing a church and a
small congregation of huts, I thought might have
been been procured, but I received the usual answer, "No hai," and I was obliged to send out for

a live sheep. I then took a siesta, and it was late in the evening before the carriage and the party arrived. They had stopped at a hut a few leagues from San Luis, and had afterwards broken the pole of the carriage, which had delayed them several hours. After supper I thought that the weather looked very wild, and I therefore got into the four-wheeled carriage to sleep, and one of the party was close to me in the two-wheeled one. The nine peons were scattered about the ground. Two of our party slept under the carriage, and the rest on the ground in different places. About midnight we were awakened by a most sudden and violent whirlwind, which blew several of the party's clothes away, and they were afterwards found in the river. There was so much dust that we could scarcely breathe, and all was utter darkness until the lightning suddenly flashed over our heads : the thunder was unusually loud, and down came a deluge of rain. The wind, which was what is termed a Pampero, was now a dreadful hurricane, and I expected every moment that it would overturn the carriage. I sat up and looked around me, and in my life I never saw so much of the sublime and of the ridiculous mixed together. While the elements were

raging, and the thunder was cracking and roaring
immediately above us, the lightning would for an
instant change the darkness to the light of day. In
these flashes I saw our party, who were all hallooing
one to another, in the most ludicrous situations.
Some were lying in bed afraid to sit up, and holding their ponchos and clothes, which were trying to
escape from them—some who had lost their clothes
were running half-naked towards the post-room—
others had lost their way, and were standing against
a dead wall, not knowing where to go. A French
Colonel, who had travelled in the carriage from
Mendoza, was lying on a stretcher made of a bullock's hide, grasping his clothes, which were now
wet through, and vociferating at his cowardly servant, who, instead of assisting him, was standing
about ten yards from him crossing himself. In vain
did he call him in Spanish every sort of " animal :"
the fellow, who had literally been approaching
his master, was rivetted to the ground by the unexpected sound of the church-bell, which, from the
violence of the hurricane, occasionally gave a solitary toll. The rain beat so violently into the two-
wheeled carriage, and it shook so terribly, that its
inmate could bear it no longer, and ran through

the rain. At last they all got into the post-room, and as I looked out of the window, I saw them all crowded together peeping over each other's heads at the door.

In the morning they found what they had lost, and the peons and the whole party looked very uncomfortable. Many of the peons had lain on the ground the whole time, and they were of course covered with the mud which had been formed by the dust and rain. The peons and the people told us they had never seen such a storm and pampero before in their lives.

The carriage was late in starting, and the sun was already up, when the French Colonel and I agreed to make a call on the priest. He was dressed in a dirty-white serge gown, tied round his body, with a rope to whip himself with; he was really not more than four feet and a half high, and yet weighed more than any of our party; his neck was as thick as a bullock's, and he had not been shaved for several days. In his room, which had no window, were two or three old books, covered with dust, and a little crucifix affixed to the wall. I asked him if it was he who had tolled the bell during the storm; he said, Oh, no! that he had ridden a number of leagues the day before, and had slept so sound, that he had not heard it, and had only just been informed of it.

On account of the party's clothes being so wet, we lost a great deal of time, and it was seven o'clock before we started. The two carriages went by the road, but the post-master told a little Gaucho to take me by a nearer cut. I followed this little child, who was not more than eight years old, for many leagues. He rode like the wind, and amused me extremely by a number of very entertaining stories which he told me. At last it began to rain, and the little boy said, " Quien sabe" if ever he should find out the post, for that he had never before come that way. It was no use stopping, and as I galloped along, I made the child tell me the directions which the post-master had given to him, but I could make nothing at all of them. One would have thought by the child's description, that it was a mountainous country we were crossing, for he talked of hills and valleys which I could not see; but the Gauchos do divide their plains into ups and downs, which no one can distinguish but themselves. At last the child exclaimed, that he could see a " Cristiano" driving some horses, and when we came to this man, he told us where the post was.

I found the horses at the post in the corral, and the post-master, whose house I had several times slept at, gave me a horse with a galope largo (a long gallop), and a very handsome Gaucho as a guide. I had a long conversation with this man as I galloped along, and I found him a very noble-minded fellow. He was very desirous to hear about the troops which the government of Mendoza had sent to reinstate the governor of San Juan, who had just been deposed by a revolution. The Gaucho was very indignant at this interference; and as we rode along, he explained to me, with a great deal of fine action, what was evident enough, —that the Province of San Juan was as free to elect its governor as the Province of Mendoza, and that Mendoza had no right to force upon San Juan a governor that the people did not approve of. He then talked of the state of San Luis; but to some question which I put to him, the man replied, that he had never been at San Luis!

" Good heavens!" said I, with an astonishment which I could not conceal,—" have you never been to see San Luis?"—" Never," he replied. I asked him where he was born; he told me, in the hut close to the post; that he had never gone beyond the plains through which we were riding, and that he had never seen a town or a village. I asked him how old he was: " Quien sabe," said he. It was no use asking him any more questions ; so, occasionally looking at his particularly handsome figure and countenance, and calling to mind the manly opinions he had expressed to me on many subjects, I was thinking what people in England would say of a man who could neither read nor write, nor had ever seen three huts together, &c. &c., when the Gaucho pointed to the sky, and said, " See! there is a lion!" I started from my reverie, and strained my eyes, but to no purpose, until he shewed me at last, very high in the air, a number of large vultures, which were hovering without moving; and he told me they were there because there was a lion devouring some carcass, and that he had driven them away from it. We shortly afterwards came to a place where there was a little blood on the road, and for a moment we stopped our horses to look at it; I observed, that perhaps some person had been murdered there; the Gaucho said, "No," and pointing to some foot-marks which were near the blood, he told me that some man had fallen, that he had broken his bridle, and that, while he was standing to mend it, the blood had evidently come from the horse's mouth. I observed, that perhaps it was the *man* who was hurt, upon which the Gaucho said, "No," and pointing to some marks a few yards before him on the path, he said, " for see the horse set off at a gallop *."

The grass was shorter in this part of the province than it usually is, and it was very picturesque and curious as we went along to see bullocks' skulls lying in different directions. The skeleton of the bull's head was justly admitted by the ancients as an ornament in their architecture. In the Pampas it is often seen lying on the ground bleached by the

---

* I often amused myself by learning from the Gauchos to decipher the foot-marks of the horses, and the study was very interesting. It is quite possible to determine from these marks, whether the horses were loose, mounted, or laden with baggage; whether they were ridden by old men or by young ones, by children, or by foreigners unacquainted with the biscacheros, &c. &c.

s.

sun, with the horns upwards, and appearing as if the animal had just risen from his grave, and was moralising to the living cattle which were feeding about him.

In consequence of what this man had told me respecting his birth, &c., I asked every one of the Gauchos who rode with me from post to post, for the next six hundred miles, the same questions, and I found that the greater number of them had never seen a town, and that no one of them knew his age. When we came to the post, which is one of the richest possessions in the Pampas, I found about twenty Gauchos assembled to commence breaking in the young horses, an operation which was to be continued for many days. As the carriage was many hours behind me, I resolved to see this, and, getting a fresh horse, I rode immediately to the corral, and soon made friends with the Gauchos, who are always polite, and on horseback possess many estimable qualities, which at the door of their hut they appear to be devoid of. The corral was quite full of horses, most of which were young ones, about three and four years old. The capataz, mounted on a strong steady horse, rode into the corral, and threw his lasso

over the neck of a young horse, and dragged him
to the gate. For some time he was very unwilling
to leave his comrades, but the moment he was
forced out of the corral, his first idea was to
gallop away; however, the jerk of the lasso
checked him in a most effectual manner. The peons
now ran after him on foot, and threw the lasso
over his four legs, just above the fetlocks, and,
twitching it, they pulled his legs from under him so
suddenly, that I really thought the fall he got
had killed him. In an instant a Gaucho was
seated upon his head, and with his long knife
in a few seconds he cut off the whole of the horse's
mane, while another cut the hair from the end of
his tail. This, they told me, is a mark that the
horse has been once mounted. They then put
a piece of hide into his mouth, to serve as a bit,
and a strong hide-halter on his head. The Gaucho,
who was to mount, arranged his spurs, which were
unusually long and sharp, and while two men held
the animal by his ears, he put on the saddle,
which he girthed extremely tight; he then caught
hold of the horse's ear, and, in an instant, vaulted
into the saddle; upon which the man, who was
holding the horse by the halter, threw the end of

it to the rider, and from that moment no one seemed to take any further notice of him. The horse instantly began to jump, in a manner which made it very difficult for the rider to keep his seat, and quite different from the kick or plunge of an English horse: however, the Gaucho's spurs soon set him going, and off he galloped, doing every thing in his power to throw his rider. Another horse was immediately brought from the corral, and so quick was the operation, that twelve Gauchos were mounted in a space which, I think, hardly exceeded an hour.

It was wonderful to see the different manner in which the different horses behaved. Some would actually scream while the Gauchos were girthing the saddle upon their backs; some would instantly lie down and roll upon it; while some would stand without being held, their legs stiff, and in unnatural directions, their necks half bent towards their tails, and looking vicious and obstinate; and I could not help thinking that I would not have mounted one of these for any reward that could be offered me, for they were invariably the most difficult to subdue.

It was now curious to look round and see the Gauchos on the horizon in different directions,

trying to bring their horses back to the corral,
which is the most difficult part of their work, for
the poor creatures had been so scared there that
they are unwilling to return to the place. It was
amusing to see the antics of the horses: they
were jumping and dancing in different ways,
while the right arms of the Gauchos were seen
flogging them. At last they brought the horses
back, apparently completely subdued and broken
in. The saddles and bridles were taken off, and
the young horses immediately trotted towards the
corral to join their companions, neighing one to
the other. Another set were now brought out,
and as the horses were kept out a very short time,
I saw about forty of them mounted. As they
returned to the corral it was interesting to see the
great contrast which the loss of the mane, and the
end of the tail, made between the horses which
had commenced their career of servitude, and
those which were still free.

The horses of the Pampas are like the common
description of Spanish horse, but rather stronger.
They are of all colours, and a great number are
pie-bald. When caught, they will always kick at
any person who goes behind them ; and it is often

with great difficulty that they can be bridled and
saddled: however, they are not vicious, and when
properly broken in, will allow the children to mount
by climbing up their tails. In mounting, it is necessary to be very quick, and previous to dismounting, it is proper to throw the bridle over one side of
the head, as the horses almost always run backwards if one attempts to hold them by the bridle
when it is over the head, as in England.

Although I rode many thousand miles in South
America, I was quite unable to learn how to select
either a good horse or an easy-going one, for by
their appearance I found it impossible to form a
judgment; indeed, I generally selected for myself
the worst-looking horses, as I sometimes fancied
that they went the best.

When first mounted, they often begin to kick
and plunge, but by giving them a loose rein, and
by spurring them, they will generally start, and
when once at their pace, they go quiet. However,
the kicking at starting is a most painful operation
to undergo, for from hard riding the back and
shoulders get so dreadfully stiff, that such sudden
and violent motion seems to dislocate the limbs.

The evening closed, but the carriages did not ap-

pear. I anxiously looked on the horizon for them,
until it became dark; I then went into the post-
room, and ordered one of the women to bring me
the roast-beef and soup which was prepared for the
party. I was quite ravenous, for I had been so occupied with the horses, that I had forgotten that I
had eaten nothing since daylight. The woman
brought me a dirty sheet four times doubled, which
she put on the little square table, then a bottle of
wine; " Have you a glass?" " No hai, senor."
" Oh, never mind," said I, putting the bottle to
my mouth. The woman returned with the beef
cut up into pieces, in a pewter dish; it was smoking, and looked very nice; and she also gave me
some bread. I instantly took out of my pocket a
clasp knife and fork. She asked me if I wanted
anything else? " No," said I, putting a piece of
the beef into my mouth; but as she was going out
of the door, I called her back, and asked her to
get me some salt. " Aqui sta, senor," said the woman, apparently recollecting herself; and opening
her right hand, she put very quietly upon the table
some salt which she had intended for me, and because some of it stuck to her hand, she scratched it

off with her fingers, and seemed resolved that I
should have every particle of it.

There was no candlestick, but, with the beef, a
little black girl about seven years old, and almost
naked, brought in a crooked, brown, tallow-candle,
which she held in her hand all the time I dined.
The little creature had gold ear-rings and a neck-
lace of red beads. I gave her a large piece of
bread, which she eat very slowly, with the most
perfect gravity of countenance. As I was dining,
I occasionally looked at her; nothing was white
but her eyes and the piece of bread in her mouth;
she was watching every mouthful I ate, and her eyes
accompanied my fork from the pewter dish to my
mouth. With her left hand she was scratching her
little woolly head, but nothing moved except her
black fingers, and she stood as still as a bronze statue.

The carriage did not arrive, so I laid my saddle
in front of the post, and slept there. It was
late in the morning before one of the peons came
to tell me that the two-wheeled carriage had broken
down in spite of all its repairs; that it was in the
middle of the plain, and that the party had been
obliged to ride, and put the baggage on post-horses,

and that they would be with me immediately. As soon as they arrived, they told me their story, and asked what was to be done with the carriage *. It was not worth more than one hundred dollars; and it would have cost more than that sum to have guarded it, and to have sent a wheel to it six hundred miles from Buenos Aires; so I condemned it to remain where it was, to be plundered of its lining by the Gauchos, and to be gazed at by the eagle and the gama,—in short, I left it to its fate.

I had been much detained by the carriages, and I was so anxious to get. to Buenos Aires without a moment's delay, that I resolved instantly to ride on by myself. Three of my party expressed a wish to accompany me, instead of riding with the carriage; so after taking from the canvass bag sufficient money for the distance, (about six hundred

---

* After the party had left one of the posts about an hour, and when they were twelve or thirteen miles from it, they saw a man galloping after the carriage, endeavouring to overtake it. They stopped, and when he came up, they found it was the master of the post-hut where they had slept. He said very civilly that they had forgotten to pay him for the eggs, and that they therefore owed him a medio, (two-pence halfpenny). They paid him the money, neither more nor less, and then galloped on, leaving the man apparently perfectly satisfied.

miles,) I left the rest for the coach, and once more careless of wheels and axles, I galloped off with a feeling of independence which was quite delightful.

We travelled sixty miles that day, not losing one moment, but riding at once to the corral, and unsaddling and saddling our own horses. The next morning one of the party was unable to proceed, so he remained at the post, and we were off before daylight. After galloping forty-five miles, another said he was so jolted that he could not go on, and he also remained at the post to be picked up by the carriage : we then continued for sixteen miles, when the other knocked up, and he really was scarcely able to crawl into the post-hut, where he remained. As I was very anxious to get to Buenos Aires, and was determined to get there as quick as my strength would allow, I rode sixty miles more that day, during which my horse fell twice with me, and I arrived at the post an hour after sunset, quite exhausted. I found nothing to eat, because the people who live at this post were bathing, so I went to another part of the river, and had a most refreshing bathe. I then spread out my saddle on the ground, for the post-room was full

of fleas and binchucas. The people had now returned from the river, and supper was preparing,
when a young Scotch gentleman I had overtaken
on the road, and who had ridden some stages with
me, asked me to come and sing with the young
ladies of the post, who he told me were very beautiful. I knew them very well, as I had passed
several times, but I was much too tired to sing
or dance: however, being fond of music, I moved
my saddle and poncho very near the party, and as
soon as I had eaten some meat I again lay down,
and as the delightful fresh air blew over my face,
I dropped off to sleep just as the niñas were singing very prettily one of the tristes of Peru, accompanied by a guitar. I had bribed the capataz to
let some horses pass the night in the corral; we
accordingly started before the sun was up, and
galloping the whole day till half an hour after sunset, we rode a hundred and twenty-three miles.
The summer's sun has a power which, to those who
have not been exposed to it, is inconceivable, and
whenever we stopped at the corral to get our horses,
the heat was so great that it was almost insup-
portable. However, all the time we galloped, the

rapid motion through the air formed a refreshing breeze. The horses were faint from the heat, and if it had not been for the sharp Gaucho spurs that I wore I should not have got on. The horses in the Pampas are always in good wind, but when the sun is hot, and the grass burnt up, they are weak, and being accustomed to follow their own inclinations, they then want to slacken their pace, or rather to stop altogether; for when mounted they have no pace between a hand-gallop and a walk, and it is therefore often absolutely necessary to spur them on for nearly half the post, or else to stand still, an indulgence which, under a burning sun, the rider feels very little inclined to grant. As they are thus galloping along, urged by the spur, it is interesting to see the groups of wild horses which one passes. The mares, which are never ridden in South America, seem not to understand what makes the poor horse carry his head so low, and look so weary. The little innocent colts come running up to meet him, and then start away frightened; while the old horses, whose white marks on the flanks and backs betray their acquaintance with the spur and saddle, walk slowly away for some distance, and then breaking into a trot, as they seek their safety, snort and look behind them, first with one eye, then with the other, turning their nose from right to left, and carrying their long tails high in the air. As soon as the poor horse reaches the post he is often quite exhausted; he is as wet as if he had come out of a river, and his sides are often bleeding violently; but the life he leads is so healthy, his constitution is so perfectly sound, and his food is so simple, that he never has those inflammatory attacks which kill so many of our pampered horses in England. It certainly sounds cruel to spur a horse as violently as it is sometimes necessary to do in the Pampas, and so in fact it is, yet there is something to be said in excuse for it; if he is worn out and exhausted, his rider also is—he is not goaded on for an idle purpose, but he is carrying a man on business, and for the service of man he was created. Supposing him to be ever so tired, still he has his liberty when he reaches the goal, and if he is cunning, a very long time may elapse before he is caught again; and in the mean while the whole country affords him food, liberty, health, and enjoyment; and the work he has occasionally performed, and the sufferings he has endured, may perhaps teach him to appreciate the wild plains in which he was born. He may have suffered occasionally from the spur, but how different is his life from that of the poor post-horse in England, whose work increases with his food,—who is daily led in blinkers to the collar, and who knows nothing of creation, but the dusty road on which he travels, and the rack and manger of a close-heated stable.

The country through which we rode this day was covered with locusts of a very beautiful colour : they were walking along the road so thick that the ground was completely covered—some were hurrying one way and some another, but the two sets were on different sides of the road like people in the City (of London). At one post these locusts were in such numbers, that the poor woman, in despair, was sweeping them away with a broom, and they swarmed in crowds up my horse's legs. A little girl had given me some water, and I put my straw hat on the ground while I sat down to drink, and with feelings of very great pleasure I was looking at the mug, which was an English one, and on which was inscribed—

> No power on Earth
> Can make us rue,
> If England to her-
> Self proves true—

when I saw my hat literally covered with the locusts biting the straw. As soon as I took it up, these parti-coloured creatures hopped off like harlequins. The number of them is quite incredible, and they would be a most serious enemy to any individual who should attempt to cultivate a solitary farm in the Pampas—although a large population and general cultivation might perhaps keep them away.

We arrived late and very tired at the post, having ridden one hundred and twenty-three miles, and found the master Don Juan —— very busy, providing supper for a priest, who had just arrived in a carriage; the water was extremely bad, and I began to think I should fare badly, when the priest asked me to partake of his supper, which was now smoking on the table. He had some good water in bottles, and we had a roasted lamb before us.

The priest ate the heart, and seemed to enjoy his repast as much as I did. He was silent, but very kind, and occasionally nodded at the dish and said to me, " Come bien!" (Eat well.) After the lamb he brought out a box of sweetmeats, and he then put his hand up the large loose sleeve of his white serge gown and pulled out some cigars.

Next morning at day-break we started. The French Colonel's servant now began to complain, and after riding one hundred miles I saw no more of him, as he and the Scotch gentleman who had accompanied me stopped at sunset. I rode on about twenty miles, and the next day I rode one hundred and twenty miles, and reached Buenos Aires about two hours after sunset.

A FEW GENERAL OBSERVATIONS RESPECT-
ING THE WORKING OF MINES IN
SOUTH AMERICA.

When one reflects upon the immense riches which have proceeded from some mines, and the large sums of money which have been lost in others, it is evident that the inspection of a mine with a view of immediately working it with a large capital, is in any country an important and difficult duty. There are, perhaps, few subjects which require more deliberate and dispassionate consideration; for to be too sanguine, or to be too timid, are faults which it is easy to commit. In the former case, one builds upon hopes which are never to be realised; in the latter, one loses a prize which energy and enterprise might have secured; and the passions of the mind are never more eager to mislead the judgment, than when the object to be considered is the acquisition of what are termed the " precious metals."

But if this is the case in civilized countries,

T

where experience has recorded many valuable data,
where the lode to be inspected may be compared with those which are flourishing and with
those which have failed, where operations may
be commenced with a cautious step, where the
windlass may be succeeded by the whims, and the
whims by the steam-engine, how much more difficult is the task when the lode is in a foreign country, destitute of resources, experience, and population, and when as a stranger one is led over a
series of wild, barren mountains, to a desert spot, at
once to determine whether the mine is to be accepted or not.   As this has been my situation, I
will enture to make a few imp erfect observations
on the subject.

The first object which draws the attention to a
lode (which is a ramified crack or fissure in which
ores with other substances are embedded), is its
positive value or contents, and this value has lately
been estimated in England merely from the inspection and assay of a piece of the ore; but of course
this judgment is altogether erroneous, for a large
lode of a moderate assay may be more valuable
than a small lode of rich ores or assay, and an

extraordinary rich lode may be too small to be worth the expense of working, while a very large poor lode may be worked with profit.

But, besides these observations, the physical character of the lode must be considered; for the fissure is seldom filled with ore, it contains also quartz, mundic,* &c. &c. and is occasionally a strong box which contains no riches at all.

It is therefore evident, that besides the size of the lode, and the assay, the average quantity of ore it contains is also to be considered; because a large lode, with an occasional bunch (as it is termed) of rich ores, may not be so worthy of working as a smaller lode with a number of bunches of poorer ores. There is also another material question,— whether the lode is getting richer or poorer as it dips? For a large lode, with rich assay, and frequent bunches, but *diminishing* in value, may be calculated at less value than a smaller lode with poorer assay, less frequent bunches, but *increasing* in value.

There are many other considerations, but the above, perhaps, will satisfy those who have not

* The Cornish term for sulphurets of arsenic, iron, &c.

reflected on the subject, that the abstract value of a lode in *America* can in no way be determined by the assay of a piece of ore in *England*, particularly when it is known that specimens of ores are often sent from South America as samples of lodes from which they never were extracted. However, upon the spot a calculation may be made of the probable produce of the mine; and in Cornwall, where the expenses of the mine are known and certain, it is on this calculation that almost all the *speculation* of the enterprise depends. For the riches of lodes being subject to sudden variations, they may increase or diminish in a surprising degree; still the calculation rests in Cornwall upon as fair a basis as those which are made on the duration of human life, or the insurance of ships, &c. &c. But in South America the case is widely different; for besides the value of the produce of the mine, it is necessary to determine what will be the probable expense of working it, in order to weigh or compare the one with the other; and the absolute necessity of this, which is always done in mining, farming *,

---

* No one would venture to say how much an unknown estate is worth per acre, merely from an inspection of a box of earth;

and other speculations in England, is particularly obvious—for instance, in the provinces of Rio de la Plata ; for as one there rides over many hundred miles of rich land, which is unowned, and almost unknown, one cannot but reflect, that while, from want of population, industry, &c., such riches are lying on the *surface* unvalued, considerable difficulties would necessarily oppose the extraction of wealth from the bowels of the earth, by labour and machinery ; and these difficulties, in many parts of the provinces, would be so great, that it might satisfactorily be proved that the silver extracted from such mines would not be worth its weight in iron by the time it reached England ; while the iron which was sent from England would cost more than its weight in silver by the time it reached the mine.

The following is a rough memorandum of some of the difficulties, physical, moral, and political,

because the object of farming being to make the receipts exceed the expenditure, it may happen (from its particular situation for manure, markets, &c.) that bad land is worth more per acre than good land.

which would probably obstruct the working of mines in the provinces of Rio de la Plata by an English association.

PHYSICAL.

1. The great distances which separate the mines from their supplies of men, tools, materials, provisions, &c., and which separate one mine from another; badness of the roads; danger in passing the laderos; torrents and rivers without bridges, and often impassable; the locality of the mines, which are generally situated among lofty and barren mountains, without resources or supplies;— the above would require expensive disbursements, and would often cause a great delay, which, in mercantile operations, is a loss of money.

2. The dryness of the climate, which affords no water for machinery, or for washing the ores; but little even to drink; the mine itself dry, or nearly so. In consequence of the above, machinery is inapplicable, and the mines are better adapted to the limited exertions of a few people, than to the extensive operations of an English association.

3. Heat of the climate; its effects on Europeans.

4. The desolate and unprotected plains between the mines and the port at which their produce would be shipped; the distance being, upon an average, more than a thousand miles of land-carriage.

5. The poverty of the lodes, when compared with those of Mexico, Peru, or Potosi.

MORAL.

The want of population—its effects.   The general want of education, and consequently the narrow and interested views of the natives.—The richer class of people in the provinces unaccustomed to business.—The poorer class unwilling to work.—Both perfectly destitute of the idea of a contract, of punctuality, or of the value of time.—Among a few people the impossibility of obtaining open competition, or of preventing the monopoly of every article required, or the combination which would raise its price " ad libitum." The wild, plundering habits of the Gauchos—the ready absolution of the priests—the insufficiency of the laws.

The want of experience, &c. in the Commis-

sioner who has charge of the Association.—The
character, constitution, habits, and expensive wants
of the English and European workmen, ill adapted
to the country.—The experience they have gained
in Cornish copper mines inapplicable to the extraction of silver ores in South America. (See Memorandum A.) Europeans, overcome by the climate,
become indolent from possessing large independent
salaries in a country where wine and spirits are
cheap—women of the country—their characters.—
Impossibility of the distant mines being frequently
inspected, consequently the necessity of placing
confidence, and of trusting gold and silver to individuals, many of whom in England would not be
deemed persons of sufficient education for so difficult a situation. Probability that many would endeavour to perform their duty, but the certainty
that one leak, whether from inattention or otherwise, would affect the interests of the whole.

POLITICAL.

Important reasons why mines in South America,
which formerly were worked with profit, would now

ruin either Europeans or natives who should attempt to work them. (See Memorandum B.)

The instability and insufficiency of the national government of the United Provinces.—The provincial governments—their sudden revolutions.—The jealousy which exists between the Provinces and Buenos Ayres.—In spite of contracts, the governments would not allow large profits to go out of their provinces, or even to pass through them without contribution.—Individuals urged by the priests would overturn the Governor—his acts and contracts fall with him.—The junta could voluntarily retire—their responsibility has then vanished—no remedy, and no appeal.

MEMORANDUM A.

Those who propose to work a mine in Cornwall,
have the following advantages over those who
propose with the same people to work a mine in
South America :—

1. In Cornwall, previous to commencing operations, they may inspect the mine themselves, and
call any number of practical men to assist them.—
In South America they cannot do this, but must
commit this important duty to one or more individuals.

2. In Cornwall the lode is in a country whose
climate is favourable to great bodily exertion, and
the general character of which is industry ; but in
South America the climate and excessive heat are
unfavourable to great bodily exertion, and the
general character of the country is indolence.

3. In Cornwall the miners are subjected to a
code of most admirable local regulations, which

encourage competition and industry, and leave the idle to starve—in South America, the miners are away from the force of all these regulations, and a high, fixed salary, with cheap wines and provisions, discourage competition and labour.

4. In Cornwall, although the miners have no theory, no schools, no books, yet, from long practice and experience, they most perfectly understand the geological construction of the country, the particular nature of the ores they seek, and the difficulties which they are likely to meet with.—In South America, the geological construction of the Andes, and the mountains in which the mines are situated, is unknown to the Cornish miner—he is unacquainted with the ores he is to seek. The muriates, carbonates, pacos, colorados, and other non-resplendent ores, are by him so unnoticed, and unvalued, that the native miner has actually to point out to him the riches of the mine he has come to improve.*

* There exists in England a natural feeling of confidence in the exertions of English workmen, but I am afraid this expectation will not be realized in South America.

The Cornish miner is, I believe, one of the best-regulated workmen in England, but like all well-regulated workmen his

5. In Cornwall, the greatest difficulties are, the subterraneous streams, which, in a humid climate and in a flat country, so influence the plan of operations, that the art of mining in Cornwall is the art of draining, not on a general principle, but adapted to the geology of the country. In South America, as it never rains at Uspallata, and seldom rains in Chili, and as the winter showers, instead of sinking into the earth, rush down the precipitous sides of the mountains in which the lodes are situated, there is but little water ; and therefore the Cornish plan of operations, and, consequently, the experience which the Cornish miner has gained, is inapplicable, for the difficulties which he has learnt to overcome do not exist;

attention has been directed to a particular object, and in proportion as he is intelligent upon that point he is ignorant of all others.

By a division of labour, which is now so well understood in England, we have goldsmiths, silversmiths, tinsmiths, coppersmiths, whitesmiths, and blacksmiths, who are all ignorant of each other's trades ; and if this is the case, why should a man whose life has been spent in working copper ores, be supposed able to search in any country for silver ores ? There is certainly a much greater difference and variety between the ores than there is between the metals.

while others oppose him which he has never been accustomed to meet.

6. In Cornwall, to drain the mines, steam-engines can be procured at a short notice, and if, for any particular object, a large body of men are required for a few days, they can always be had; also whatever tools, wood, iron, rope, &c., may be required, can be obtained with a facility and punctuality known only in England. In South America, from the absence of water, the overpowering force of steam is unnecessary, inapplicable, and its great advantage is unattainable. In case of unforeseen difficulties requiring for a few days the assistance of a large body of extra labourers, it would be absolutely impossible to obtain them. Tools, iron, and materials could only be procured with the greatest possible difficulty. In many situations it would be necessary to send several hundred miles for materials. The purchaser would be assailed by every endeavour and combination to defraud: they would be delivered at a great expense of time and money; and in a country in which contracts are not understood, and time is of no value, there would be the most serious delays and disappointments.

7. In Cornwall, the expenses of the mine are known. The customary wages of the captains of the mines, the pay of the miners, who all work by tribute*, or by tutwork, are accurately calculated; the price of tools, iron, wood, rope, and all materials is known, and the sale of the ores by public auction gives an immediate and certain return. In South America the expenses of each mine can never be anticipated. The wages of the English captains and miners are very high ; every article, if purchased a thousand times, would be the subject of a new bargain, and materials would be perhaps of double or treble cost, according to the people, and the spots from which they were to be obtained.

* Excepting the levels, which are always driven by tut-work (task-work), the mines in Cornwall are all worked by Tributers. These Tributers are the common miners, who take their pitches by public auction, at which they agree to deliver the ore fit for market for different prices, from 6*d*. to 13*s*. 4*d*. in the pound, according to the nature of the ground, the ores, &c.&c. The adventurers of the mine, therefore, are tolerably sure of their profit before the work is begun, for the Tributers pay the smith-cost, candles, powder, breaking, wheeling, and drawing. They pay men for spalling and cobbing the large rocks, for separating the prill from the dradge, and they also pay girls for bucking the ores, and boys for jigging them.

After the extraction and reduction of the ores, the processes of smelting and amalgamation, which in Cornwall are unknown, (the Cornish ores being always smelted in Wales,) would be required.

8. In Cornwall, in case it should be deemed necessary to abandon the mine, the men can be discharged; the engines can be removed; the materials can be sold by auction, and the loss is only what has actually been spent on the mine. In South America, in case the mine should be deserted, to the sum sunk in the mine is to be added, the expense of the men getting to the spot and returning, which in many cases would be very great; the construction of houses for officers and men, as also the establishments for smelting and amalgamation; the cost of engines and stores, which it would often be cheaper to abandon than to remove.

9. In Cornwall, the resources of a great mercantile country are so extensive, that public competition suppresses every sort of unjust combination, but among small communities of men this would be impossible; and without the slightest intention to blame any individual, I must declare, that from

the Atlantic to the Pacific, I found that Englishmen and foreigners were preparing to monopolize
every article that could be required for mining purposes; and that a large English capital, belonging
sometimes to A., and sometimes to B., was considered by a pack of people as a headless, unprotected carcass, which was a fair subject for uni-
versal " worry."

Memorandum (B).

Comparing the past and the present Value of the Mines in South America.

On the discovery of the different countries of South America, the attention of the Spaniards was immediately directed towards the acquisition of those metals which all men are so desirous to obtain. Careless of the beauty of these interesting countries, their sole object was to reach the mines; and hence it is that the history of the American mines has always been considered the best history of the country. As soon as information was obtained from the Indians of the situation of the mines, however remote, small settlements were formed there; and with no other resources or supplies than those which nature had bestowed upon the country, they commenced their labours: they obtained their reward, and the arrival of the precious metals in Europe was hailed as the produce of intrepidity, industry, and science

The mode, however, in which these riches were at first obtained, forms one of the most guilty pages in the moral history of man; and the cruelties which were exercised in the American mines are a blot on the escutcheon of human nature, which can never be effaced or concealed, and which is now only to be confessed with humility and contrition. Besides the mita, or forced labour of the Indians (the particular cruelty of which it is not the present object to describe), the whole system was one of extortion and oppression *. The miners were barely sheltered from the weather; the use of all spirits was forbidden; their food was coarse, and the weighty tools which were placed in their hands

* Those who formerly worked the South-American mines have been accused of *ignorance,* in having brought ore and water from the mine on the backs of men. If the Indians employed had received English wages and English comforts, and had carried the small quantity which in England would be called a load, the *ignorance* of their masters would have been great indeed. But the case was very different. The Indian Apires were beasts of burden, who carried very nearly the load of a mule; and their food cost but little. Their unrecorded sufferings were beyond description; and I have been assured, from unquestionable authority, that, with the loads on their backs, many of them threw themselves down the mine, to end a life of misery and anguish.

were in themselves emblems of the ignorance, cruelty, and avarice of their masters.

However, there is no situation of misery or suffering to which the mind and body of man cannot be enured. The miner by degrees became accustomed to his labour and his tools; the slave, toiling under his load, ceased to complain; the cry of the sufferers became gradually silent, and in a short time no sound issued from the gloomy chamber of the mine but the occasional explosion of powder, the ringing blow of the hammer, and the faint whistle of the slaves, who thus informed the overseer that they had reached those points of the shafts at which, by law, they were allowed to rest.

The mine was said to have assumed a prosperous appearance, and men were talking aloud of the flourishing state of the South-American colonies, and of the inexhaustible riches of the mines, when the spell was gradually broken. The revolution at last broke out, and, as if by magic, the miner found himself in the plain surrounded by his countrymen, marching forward in support of liberty, and lending his arm to exterminate from la Patria the oppressors who were now trembling before them.

All the poor mines in South America from this moment were deserted, and the country was for many years in a state of warfare which it is not necessary to describe ; but as soon as the victory was gained, and independence gradually established, one of the first acts to which many people had recourse, was the working of the deserted mines, from which they naturally expected again to obtain wealth. Several of the miners had been killed in the wars, and others, wearing the spurs and poncho of the Gaucho, enjoyed a life of wild and unrestrained liberty. There were some, however, who voluntarily returned to the profession in which they had been trained, and were willing again to embrace a life whose hardships had become habitual ; but the forced labour of the Indian was now wanting ; and although this system of cruelty had been long abolished in many parts of South America, yet its existence in some places, and the unjust encouragement which the Spaniards had given to mining, in exclusion of every other branch of industry, had, up to the period of the Revolution, greatly assisted the working of the mines.

Operations were, however, recommenced at almost all the old mines. They were all tried ; but, generally speaking, they were all abandoned, because they did not pay, and with little inquiry into the cause, the reason assigned was, the want of intelligence and capital; and people, frustrated in this object, and incapable of contending with the difficulties which impeded any step towards civilization in the insulated, remote, and almost impracticable situations in which they often found themselves, yielded to the habits of indolence in which they still exist.

If the above rough and imperfect description of the mines of South America is deemed correct in its general features, it will account for a phenomenon which, in visiting several deserted mines, I was for a long time totally unable to comprehend.

In many places we found lodes worked to considerable depths, but the lode so small, and the assay so poor, that the constant remark of the Cornish captains who accompanied me was, " that there must have been something got out of the mine which they could not see, or else it could never have paid."   Besides this, the country was barren, and there were often many other local disadvanages: still, however, it was evident to me that these mines somehow or other MUST have paid, or else they would not have been worked; and in spite of the disadvantages which were *before my eyes,* the natural conclusion was, that if they had once paid, they might surely pay again.

However, as soon as I afterwards saw a few of the miners at work, the problem was solved.

The miners who are now in Chili, though toiling in the path of their early days, have probably relaxed a little from the discipline of the Spaniards; but the extraordinary manner in which they still work, or rather slave, is almost incredible. The contrast between their lives and the ease and independence of the rest of the inhabitants of the country, naturally leads the mind to reflect on the sad history of the South American mines; and this history, in my humble opinion, sufficiently accounts for, 1st, the impossibility which now exists of getting more miners; and, 2ndly, for the important truths, that the American mines have positively fallen in value since the country has been free, because the contents or produce of the mines are still the same, while the value of labour, &c., has ne-

cessarily increased; and therefore that, far from
being able to get a *greater* profit from these mines
than was extracted by the Spaniards, it would be
impossible now to draw from them what they formerly repaid, and that many of them must continue deserted, for the evident reason, that poor
mines, as well as poor land, may be made productive by a system of cruelty and tyranny, when
under a free government they must lie inactive and
barren.

CONCLUSION.

HAVING now completed a very rough and defective sketch of the Pampas, &c., and some of the provinces of the Rio Plata and of the governments and habits of the people, it is natural to consider how powerful this country must necessarily become, when, animated by a large population, enriched by the industry and intelligence of man, and protected by the integrity and power of well-constituted governments, it takes that rank in the civilized world which is due to its climate and soil; and as, in Nature's great system of succession, " nations and empires rise and fall, flourish and decay," it is possible that this country, availing itself of the experience of past ages, may become the theatre of nobler actions than any of the nations of the Old World, whose obscure march towards civilization was without a precedent to guide them, or a beacon to warn them of their dangers. And far from being jealous of the superior strength and energy which a young country may attain, it is pleasing to anticipate the prosperity which may await it, and to indulge a hope that its young arm may assert the dignity and the honour of human nature; that it may liberate the slave, and against every threat or danger support freedom, when the infirmities of an *older nation* may have rendered her incapable of the task.

But between this moral and political eminence which the Pampas and the provinces of Rio Plata may attain, and their present state, there is a distance which is evident to every one, though no man can calculate the time which will be requisite to pass it. The difficulties to be encountered must necessarily be great, and it is not an improper or a useless subject of speculation, to consider what some of these difficulties may be.

The great desideratum of these countries is population; for until there is a certain proportion of inhabitants, the provisions of life must necessarily be easily obtained, and people will remain indolent, until necessity drives them to exertion. The overplus population of the Old World will undoubtedly flow towards these countries, bringing with it different habits, languages, and customs. The points at which the emigrants settle will depend upon the produce which they are best fitted for obtaining, and the governments of the different provinces must become more or less powerful in proportion to the success of these people. Some will rapidly rise, while others will be left for some time in the wretched state of poverty and inactivity in which they now exist; and the laws and regulations which govern the one will be insufficient, inapplicable, or contrary to the interests of the others. As the provinces become more vigorous, it will probably be found that the situations of many of the present capitals must unavoidably be changed. For instance, the maritime province of Buenos Aires already requires a harbour; and it is easy to foresee, that when commerce establishes its residence at the new port, the government must follow.

The language, religion, habits, and occupations of the different provinces will of course be influenced and effected by the quantity of foreign settlers, and the laws MUST vary with the exigencies which require them. The provinces, as they become powerful, will naturally desire to be independent; and the possibility of their being all governed from Buenos Aires will rapidly diminish.

During these or similar events, the provinces of the Rio Plata must necessarily be in a troubled and unsettled state. The national government, thwarted in its plans, deserted sometimes by one province, and sometimes opposed by another, must often, unavoidably, act contrary to the interests of those plans it may have suggested ; while the provincial governments must often suddenly be overturned, be annihilated and remodelled, until prosperity has afforded to society the liberal principles of a good education, which, with time and experience, will at last constitute governments practically suited to the country.

If the state of the provinces of Rio Plata has been correctly sketched, and if the above should be a fair statement of some of the probable difficulties which these provinces will experience in their progress towards civilization, there are two questions to be considered, which are very material to the interests of many individuals in our country.

1st. *Is it adviseable for those who are in reduced circumstances in England to migrate to these provinces?*

2nd. *Is it prudent for those of large capital to embark their money there in any permanent establishment or speculation ?*

My humble opinion on these two important questions is shortly as follows:—

A poor individual, or a poor family, or a congregation of poor families, coming from England to these provinces, will instantly be relieved from that part of their sufferings which proceeded from absolute want of food, for they will arrive at a place where coarse beef is cheap. Artizans will obtain good wages in the town of Buenos Aires; but as English peasants are not fitted to perform any part of the Gaucho's labour, they will not receive from them more than their board.

Now, at Buenos Aires, artizans will find provisions very dear, and although they receive more money than in England, they will not be able to live there so well. The lodgings, which are always unfurnished, are shockingly dirty, filled with all sorts of vermin; and, after all, they are extremely dear. Beef is sold in such a mangled state, that when the Cornish miners first arrived, they often returned from the butchers' carts without buying the meat, being unable to make up their minds to eat it. The fowls at Buenos Aires are also very bad, for they feed upon raw meat; occasionally I have seen them hopping out of the carcass of a dead horse; and we all fancied that the eggs tasted of beef. The pigs are also carnivorous. Raw beef is cheap, but fuel*, pepper, salt, bread, water, &c., are all so exorbitantly dear, that the meat when cooked positively becomes expensive; and every article of clothing is eighty per cent. dearer than in England.

The society of the lower class of English and Irish at Buenos Aires is very bad, and their constitutions are evidently impaired by drinking, and by the heat of the climate, while their morals and characters are much degraded. Away from the religious and moral example of their own country, and out of sight of their own friends and relations, they rapidly sink into habits of carelessness and dissipation, which are but too evident to those who come fresh from England; and it is really too true, that all the British emigrants at Buenos Aires are sickly in their appearance, dirty in their dress, and

* The coals which are used come from Newcastle; and almost all the potatoes from Falmouth.

disreputable in their behaviour. A poor person
with a young family should therefore pause before
he brings them into such society; for it is surely
better that his children, until they arrive at an
age to work, should occasionally be in want in
England, than that their constitutions should be
impaired, and those principles ruined, which induce every religious and honest man in England to
labour with cheerfulness, and to return from his
work with a healthy body and a contented mind.

A single man may imagine that he is able to
resist the effects of bad society; that he would
enjoy the climate and freedom of the country, and
by attention, save up a sum of money to return to
England,—but he would find many unexpected
difficulties.

The principal one to a working man is the climate, which in summer is so dreadfully hot that
his constitution is unable to stand against it, and
with every inclination to work he finds that his
strength fails him, and that he is overpowered by
a debility before unknown to him. He would then
wish himself back in England, and his absence
from his friends, and being unable to work, would

make him discontented with a life which hangs heavy upon his hands, and which becomes more cheerless, because, unless he has a large sum of money, to pay for his passage, he sees that he is unable to return.

The above observations are not altogether theoretical. I particularly observed the unexpected effect which the climate had upon many English companies*, and upon a large body of our English

---

* We had all sorts of English speculations in South America, some of which were really amusing. Besides many brother companies which I met with at Buenos Aires, I found a sister association of milkmaids. It had suddenly occurred to some of the younger sons of John Bull, that as there were a number of beautiful cows in the United Provinces of Rio de la Plata, a quantity of good pasture, and as the people of Buenos Aires had no butter to their bread, a Churning Company would answer admirably; and before the idea was many months old, a cargo of Scotch milkmaids were lying becalmed under the Line, on their passage to make butter at Buenos Aires. As they were panting and sighing (being from heavy rains unable to come on deck), Neptune as usual boarded the ship, and the sailors who were present say that his first observation was, that he had never found so many passengers and so few beards to shave; however, when it was explained to him, that they were not Britannia's *sons*, but Jenny Bulls, who have no beards, the old god smiled and departed. The people at Buenos Aires were thunderstruck at the unexpected arrival of so many British

miners, who were selected in Cornwall for their
good behaviour, and who arrived in the Provinces
with every inclination to maintain their character.
They saw the degraded state of the English settlers
at Buenos Aires, and of their own accord they kept
clear of them; but the cheapness of the spirits,
and the heat of the climate, were inducements to
them to drink, which they found it very difficult to
resist. As soon as the heat set in, the men were
exhausted, and complained of a " feebleness" that
they had never felt before; and this was so great,
that many of the strongest of them preferred going

milkmaids; however, private arrangements had been made,
and they, therefore, had milk before it was generally known
that they had got cows. But the difficulties which they expe-
rienced were very great : instead of leaning their heads against
patient domestic animals, they were introduced to a set of law-
less wild creatures, who looked so fierce that no young woman
who ever sat upon a three-legged stool could dare to approach,
much less to milk them!—But the Gauchos attacked the cows,
tied their legs with strips of hide, and as soon as they became
quiet, the shops of Buenos Aires were literally full of butter.
But now for the sad moral of the story :—after the difficulties
had been all conquered, it was discovered, first, that the butter
would not keep!—and secondly, that somehow or other the
Gauchos and natives of Buenos Aires - - - - - - - - liked oil
better !!

without meat to the fatigue of going through the sun to fetch it. This imbecility had its natural effect upon their minds, and they expressed their dislike of a climate in which they could make no exertions, and by which they were even exhausted while lying down or sitting still; and as soon as I determined on sending them home, they all most joyfully gave up the lucrative advantages which had induced them to come to the country, and none of them would remain, although by their agreements they might each have claimed sixty pounds instead of a passage, and might instantly have made very good contracts with the other Mining Companies; but they were all anxious to return, and I heard several of them say to each other, that " they had sooner work their fingers to the stumps in England than be gentlemen at Buenos Aires."

From the above circumstances, and many other observations which I endeavoured to make on the situations of a few English emigrants I met with in the different Provinces, I am convinced that those who have hitherto emigrated to this country, as well as those who deserted from General Whitelocke's army, have passed their days in disappointment

and regret—that the constitution of every individual has been more or less impaired—that their religious principles have altogether been destroyed—
and I therefore would sincerely advise poor people,
particularly those who have families, not to migrate
to such hot latitudes, if they have the means of supporting themselves in England.

In reply to the second question, *Whether it is
prudent to embark a large capital in any permanent
establishment or speculation in this country?*—the
Spanish South Americans have certainly become independent of the government of Spain, and this
has of course proceeded from their own positive
strength, and from the imbecility of the Spanish
government; but supposing it to have arisen from
the first cause only, still it must be admitted that
a young nation may be strong enough to gain its
independence, before it has education, wisdom, or
experience enough to know what to do with it; and
taking into consideration the peculiar political situation of the country, I must own it appears to me
that during the troubles and vicissitudes which
must unavoidably attend the progress of these provinces towards civilization, it would be imprudent

for a stranger to enter into any permanent establishment; for, ignorant of what is to happen, all he
can depend upon is, that great changes will take
place, that *he* must always be a responsible person,
while unlooked-for revolutions may cause the governments or the individuals with whom he has established himself to vanish, leaving him in the wide
plain without a remedy, and perhaps even without
a just cause of complaint. He may have treated
with a government which has ceased to exist, or
with an individual whose fortune or whose influence
may have suddenly disappeared; and be like the
person who came from England to Buenos Aires
some years ago, under the promise that he should
have a lucrative situation in the Cabildo, and who
learnt on his arrival that the Cabildo had just been
destroyed.

I can speak from my own private experience, for
I was very nearly in a similar or a worse situation.
I was furnished with letters of introduction to the
Governor of San Juan, and a copy of the then
famous Carta de Mayo, which had been published
in that province to insure to us religious toleration;
but had I not fortunately been delayed upon my

road, I should upon my arrival at San Juan have been instantly thrown into prison with the Governor who was already confined, and from the window of my dungeon I should have seen the public executioner burning the Carta de Mayo, amidst the acclamations of the people. Yet I could not have complained, for my letters of introduction and the copy of the Carta de Mayo had been sent to me with the best intention—and the Governor at San Juan had wished to give me a polite reception; but the event was a political tempest which had not been foretold.

The failure of the Rio Plata Mining Association is a serious proof of the insufficiency of the Governments of La Plata. This public association was formed in London in virtue of a Decree, &c., from the government of Buenos Aires, authorizing the formation of a Company to work the mines of the United Provinces, at the discretionary choice of the Company; and to promote this object, Reports were forwarded from the Governors of the Mining Provinces describing their Mines. Yet, on my arrival at Buenos Aires, I found that almost the whole of the mines were already sold by the

Governments to the opposition Companies, and that the Government of Buenos Aires, as well as the Governors of the Provinces, had been totally unable to fulfil the Decree. Private interests and private speculators had overpowered their act and their intention, and they had only to confess—

Tempora mutantur, et nos mutamur in illis.

THE END.

Printed in the United States
By Bookmasters